Energy Alternatives

Other books in the Current Controversies series:

The AIDS Crisis
Drug Trafficking
Iraq
Police Brutality
Women in the Military

Energy Alternatives

Charles P. Cozic, *Book Editor*
Matthew Polesetsky, *Book Editor*

Cover photo: FPG

Library of Congress Cataloging-in-Publication Data

Energy alternatives / Matthew Polesetsky & Charles P. Cozic, book editors.
p. cm. — (Current controversies)
Includes bibliographical references and index.
Summary: A collection of articles offering opinions for and against such energy issues as whether fossil fuels should be replaced, the uses of nuclear power, alternatives to gasoline-powered cars, and the need for a national energy policy.
ISBN 0-89908-577-6 (lib. bdg. : alk. paper) — ISBN 0-89908-583-0 (pbk. : alk. paper)
1. Renewable energy sources—Juvenile literature. 2. Fossil fuels—Juvenile literature. [1. Renewable energy sources. 2. Fossil fuels.] I. Polesetsky, Matthew, 1968- . II. Cozic, Charles P., 1957- . III. Series.
TJ808.2.E63 1991
333.79—dc20 91-24387

Printed on recycled paper

Contents

Foreword

By definition, controversies are "discussions of questions in which opposing opinions clash" (*Webster's Twentieth Century Dictionary Unabridged*). Few would deny that controversies are a pervasive part of the human condition and exist on virtually every level of human enterprise. Controversies transpire between individuals and among groups, within nations and between nations. Controversies supply the grist necessary for progress by providing challenges and challengers to the status quo. They also create atmospheres where strife and warfare can flourish. A world without controversies would be a peaceful world; but it also would be, by and large, static and prosaic.

The Series' Purpose

The purpose of the Current Controversies series is to explore many of the social, political, and economic controversies dominating the national and international scenes today. Titles selected for inclusion in the series are highly focused and specific. For example, from the larger category of criminal justice, Current Controversies deals with specific topics such as police brutality, gun control, white collar crime, and others. The debates in Current Controversies also are presented in a useful, timeless fashion. Articles and book excerpts included in each title are selected if they contribute valuable, long-range ideas to the overall debate. And wherever possible, current information is enhanced with historical documents and other relevant materials. Thus, while individual titles are current in focus, every effort is made to ensure that they will not become quickly outdated. Books in the Current Controversies series will remain important resources for librarians, teachers, and students for many years.

In addition to keeping the titles focused and specific, great care is taken in the editorial format of each book in the series. Book introductions and chapter prefaces are offered to provide background material for readers. Chapters are organized around several key questions that are answered with diverse opinions representing all points on the political spectrum. Materials in each chapter include opinions in which authors clearly disagree as well as alternative opinions in which authors may agree on a broader issue but disagree on the possible solutions. In this way, the content of each volume in Current Controversies mirrors the mosaic of opinions encountered in society. Readers will quickly realize that there are many viable answers to these complex issues. By questioning each author's conclusions, students and casual readers can begin to develop the critical thinking skills so important to evaluating opinionated material.

Current Controversies is also ideal for controlled research. Each anthology in the series is composed of primary sources taken from a wide gamut of informational categories including periodicals, newspapers, books, United States and foreign government documents, and the publications of private and public organizations.

Readers will find factual support for reports, debates, and research papers covering all areas of important issues. In addition, an annotated table of contents, an index, a book and periodical bibliography, and a list of organizations to contact are included in each book to expedite further research.

Perhaps more than ever before in history, people are confronted with diverse and contradictory information. During the Persian Gulf War, for example, the public was not only treated to minute-to-minute coverage of the war, it was also inundated with critiques of the coverage and countless analyses of the factors motivating U.S. involvement. Being able to sort through the plethora of opinions accompanying today's major issues, and to draw one's own conclusions, can be a complicated and frustrating struggle. It is the editors' hope that Current Controversies will help readers with this struggle.

Introduction

On September 4, 1990, Eric Raymond landed an odd-looking airplane in a farmer's field near the coast of North Carolina. The landing marked the first U.S. coast-to-coast flight in a solar-powered aircraft. Raymond, who had begun his journey one month and two days earlier in the Southern California desert, had equipped his custom-built plane, *Sun-Seeker*, with seven hundred solar cells to convert sunlight directly into electricity. The electricity powered a three-horsepower motor and propeller used during takeoffs, landings, and wind turbulence. Aside from these times, the *Sun-Seeker* glided across America on updrafts of hot air.

This accomplishment represents one of the latest applications of solar energy, just one of many energy alternatives known as renewable energy sources. Others include wind power, hydroelectric power, ocean wave power, and geothermal energy, which is heat from beneath the earth's surface. The number of renewable energy projects around the world are increasing. For example, an ocean wave energy plant on the coast of Ireland is harnessing the power of the Atlantic Ocean's tides. Similar plants will be constructed in Japan.

In the U.S., particularly in Southern California, many renewable energy installations now generate electricity. Near Palm Springs, four thousand windmills, technically known as turbines, dot the landscape, producing electricity from wind for a major utility company. Statewide, approximately sixteen thousand wind turbines generate as much energy per year as a medium-sized nuclear plant. Solar power is thriving, too, in California's Mojave Desert. At Kramer Junction, four hundred acres of curved mirrors convert solar rays into electricity for LUZ International, a Los Angeles-based solar energy firm. As it works to make solar power even more inexpensive, LUZ expects to meet the energy needs of approximately one million people by 1994.

Many energy researchers are confident that renewable energy sources will eventually replace electricity generated by oil, coal, and nuclear power. Renewable energy provides 10 percent of the nation's electricity, approximately half as much as nuclear power. According to the Union of Concerned Scientists environmental group, renewable energy can potentially supply half of America's energy requirements by the year 2020.

However, other energy experts are less optimistic about the future of renewable energy. They cite the fact that interest in renewable energy, particularly solar power, waned during the 1980s, when low gasoline prices and a worldwide glut of oil made fossil fuels much more cost-effective than renewable energy. Many oil companies which had invested money in solar energy projects during the oil shortages of the 1970s withdrew support a decade later as oil prices dipped into the $15-to-$20-per-barrel range and the demand for new energy sources declined. In 1989, the Atlantic Richfield oil com-

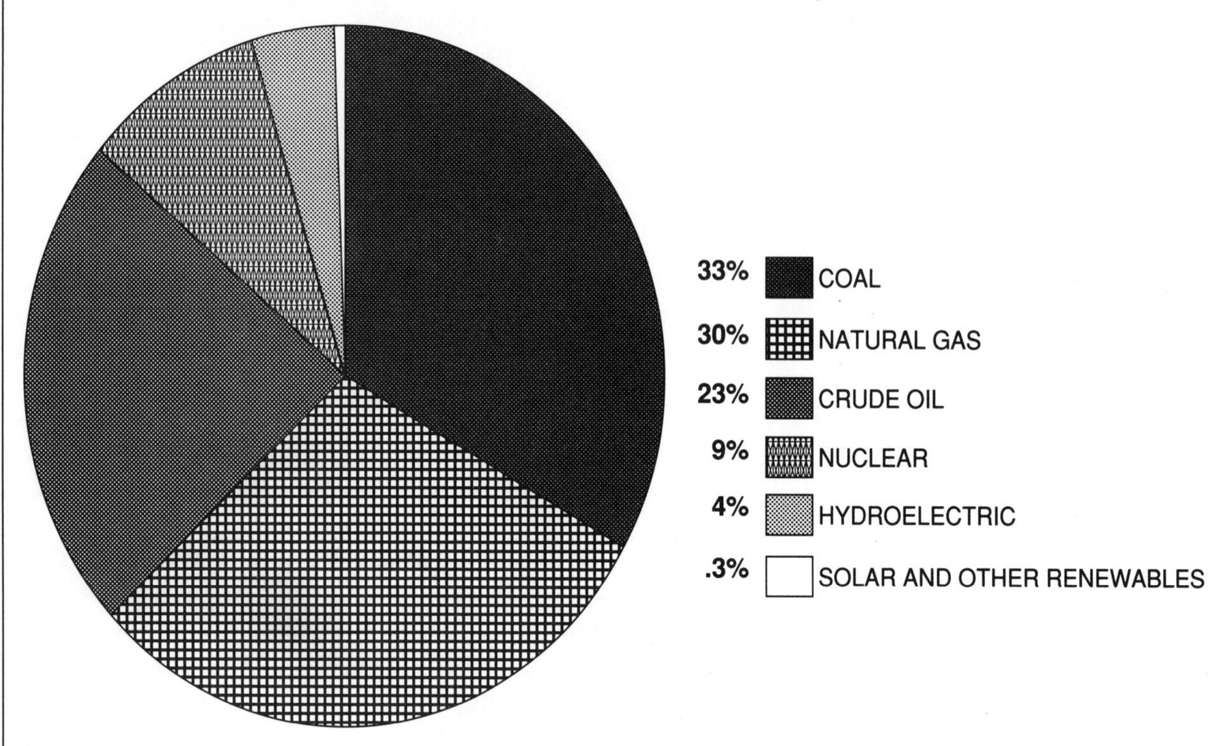

U.S. Electricity Generated by Renewable Energy Sources, 1990

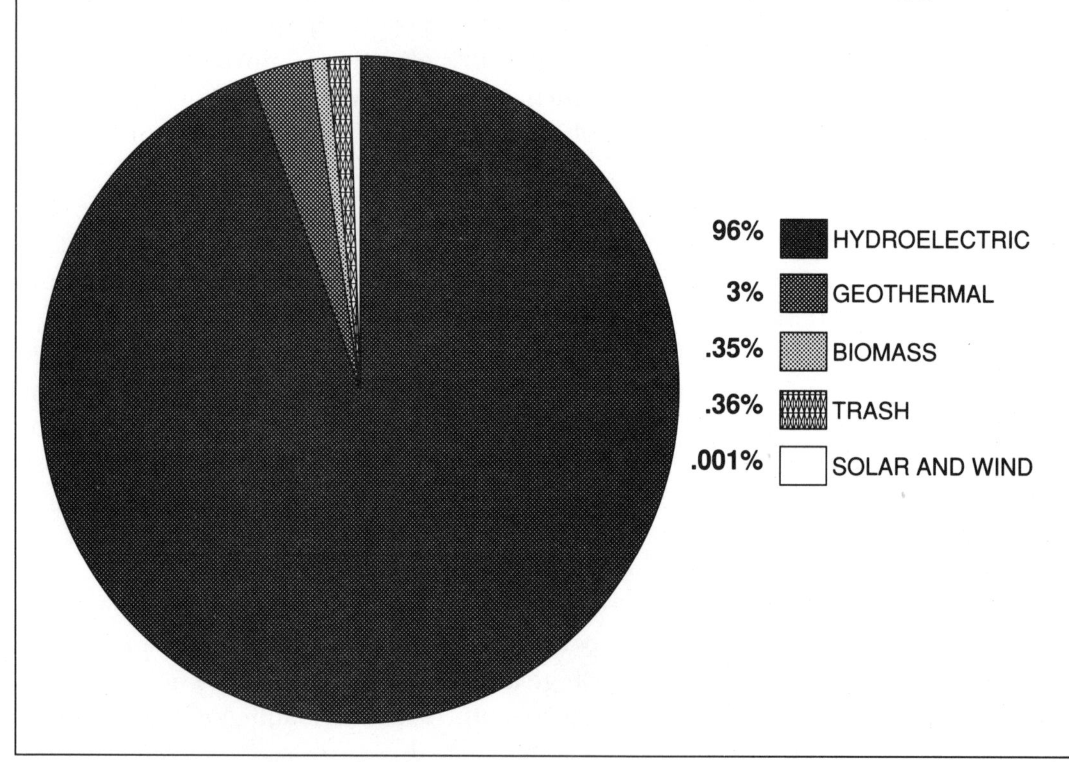

pany, commonly known as ARCO, ended a twelve-year effort to become the world's largest maker of photovoltaic solar panels, which use the same technology as the *Sun-Seeker* airplane. ARCO sold its solar subsidiary to a foreign corporation because it questioned the future need for solar power.

U.S. banks also became increasingly reluctant to support renewable energy production. In the late 1980s, many small energy companies that sought financing for renewable energy projects were rejected by banks. Said the chairman of one such company, "We couldn't get to first base with any U.S. banks. They didn't want to take the risk." Corporations such as General Electric and Westinghouse Electric also lost interest in the renewable energy business. In 1989, thirty-five of more than one hundred U.S. alternative energy companies profiled in a study three years earlier had ceased operating. The federal government also reduced much of its federal funding for renewable energy research during the 1980s. Deprived of financial assistance from both private and public sources, many renewable energy companies could not stay in business.

Both advocates and opponents agree that renewable energy sources will be viable alternatives only if they can become economically competitive with fossil fuels, or when exorbitant oil prices force consumers and corporations to reconsider renewable energy for their energy needs. Thus, the future of renewable energy, much like any innovation, depends on technological and economic forces. It remains to be seen whether renewable energy will revolutionize the way Americans use energy. The authors of the viewpoints in *Energy Alternatives* discuss the capabilities of renewable energy and other energy sources.

Chapter 1: Preface

Should the U.S. Decrease Its Use of Fossil Fuels?

Fossil fuels are the remains of ancient animal and plant matter that have been transformed by the earth's natural processes. Because fossil fuels are usually located deep within the earth, mining or drilling are required to bring them to the surface. When burned, fossil fuels emit a great deal of energy which can be used to power cars, heat homes, or generate electricity.

America relies heavily on fossil fuels to meet its energy needs. Together these fuels supply nearly 90 percent of America's energy requirements. Yet the current deposits of these fuels are finite, and once they are depleted they cannot be replaced. This fact has led some experts to suggest that the U.S. should decrease its dependence on fossil fuels. Others, however, maintain that fossil fuels provide an efficient, inexpensive source of energy that the nation cannot do without.

Those who favor decreasing the use of fossil fuels point out that such fuels are the primary source of pollution in the U.S. When burned, fossil fuels produce carbon dioxide, the primary gas alleged to contribute to global warming. According to many experts, carbon dioxide emitted from the burning of fossil fuels accumulates in the atmosphere, creating a heat-trapping blanket around the earth. The result, these experts believe, will warm the planet beyond its normal temperature, leading to floods, crop failures, and the extinction of species.

Global warming is not the only environmental threat posed by the use of fossil fuels, many experts contend. These people warn that pollution from fossil fuels acidifies the moisture in the air to create acid rain. Acid rain harms trees, fish, and crops and can contaminate drinking water. Smog, or air pollution, is another consequence of burning fossil fuels. Smog endangers the health of Americans by increasing or aggravating respiratory problems. Finally, the transportation of oil poses environmental risks as well. Oil spills from damaged tankers can kill birds, otters, and other wildlife, as seen in the disastrous 1989 *Valdez* spill in Alaska, in which eleven million barrels of oil poured into the waters of Prince William Sound.

U.S. dependence on foreign oil is another reason to decrease its use, many analysts argue. In 1990, the United States depended on oil imports for about 47 percent of its oil, up from about 32 percent in 1985.

Other analysts, however, believe the advantages of fossil fuels far outweigh their disadvantages. Fossil fuels are a cheap and plentiful source of energy, these analysts maintain. "There is no present substitute to fuel the engines of advanced societies," according to columnist Woody West. If the U.S. attempted to switch to alternative forms of energy, many experts contend, it would wind up paying much more for less energy and suffer a serious decline in its living standard.

These experts also argue that the environmental problems caused by fossil fuels are exaggerated. For instance, according to fossil fuel advocates, there is only meager evidence supporting the theory of global warming. This evidence does not justify switching from fossil fuels to expensive alternative sources of energy. Also, "clean coal" technologies, which remove pollutants from coal, have dramatically reduced the problem of acid rain. Experts who support the use of fossil fuels also believe that oil spills do not cause permanent damage and that with increased care, spills can be prevented.

Fossil fuel advocates also assert that America's dependence on foreign oil could be greatly reduced by increased use of domestic fossil fuels. For example, coal, which in 1989 produced more than half of America's electricity, is found in great abundance in the United States. In fact, these advocates point out, the United States has the largest known recoverable coal reserves of any nation in the world, enough to last the country for three centuries at current rates of use. The United States also has large untapped reserves of oil in Alaska and in offshore locations, many experts believe.

Fossil fuels play a central role in the American economy. Their impact is felt in every home, office, and factory in the country. However, some experts argue that the environmental and political costs associated with fossil fuels outweigh their benefits. Others disagree, insisting that the country can continue to use fossil fuels to maintain a high standard of living while still protecting the environment. The following chapter debates the issues surrounding the use of fossil fuels.

Should the U.S. Decrease Its Use of Fossil Fuels?

Yes: Fossil-Fuel Use Should Be Decreased

The U.S. Should Decrease Its Use of Fossil Fuels

Alexandra Allen

About the Author: *Alexandra Allen is a staff member of Greenpeace, an international environmental and antiwar organization politically involved in a variety of environmental and peace issues.*

When Saddam Hussein put forward an offer on February 15, 1991, to withdraw from Kuwait under certain conditions, President George Bush termed the offer a "cruel hoax," and repeated his call for an unconditional withdrawal. And so the soldiers resumed a war which most of the world cannot help but recognize was fought over oil—a finite resource that is itself imperiling the survival of life on the planet.

This tragic irony makes it abundantly clear that unless the energy economy of the United States and other industrialized countries changes dramatically, the world is headed for a bleak future of oil wars, oil spills, global warming and polluted air and water. But if anyone thought that President Bush's recently proposed "National Energy Strategy" (NES) would chart the course toward a sensible and sustainable energy future, it is they who have fallen prey to what can truly be called a cruel hoax.

The energy plan that President Bush announced on February 20, 1991 is draped in the language of "increasing energy and economic efficiency," "securing future energy supplies," "respecting the environment" and "preserving the free market." Behind the technocratic jargon and nods toward environmental concerns lies the reality that the Bush plan makes no serious effort to wean the United States away from its fossil fuel dependency. As Peg Stevenson, director of the Greenpeace Atmosphere and Energy Campaign, says, "The plan ignores energy efficiency, continues to define energy security as access to oil and fails to embody an understanding of the environment, let alone any respect for it. The rhetoric about the free market is simply a euphemism for protecting major corporate interests in the fossil fuel and nuclear industries."

Alexandra Allen, "Bush's National Energy Policy." This article first appeared in the March 1991 issue of *Multinational Monitor*, PO Box 19405, Washington, DC 20036; subscriptions, $22/year. Reprinted with permission.

"The Bush plan makes no serious effort to wean the United States away from its fossil fuel dependency."

The development of the National Energy Strategy was initiated by Bush in June 1989, when he called on the Department of Energy (DOE) to begin a "dialogue with the American people" on energy and to develop energy policy recommendations for the nation. DOE held 18 public hearings around the country and heard from 448 witnesses, the majority of whom were from energy corporations. While many national environmental groups testified at the hearings, few local citizens' organizations participated. In some cases, witness lists were finalized by DOE before the hearing was even announced publicly. The domination of industry witnesses did not blunt the widespread support for improved energy efficiency, however. DOE officials acknowledged in an interim report on the National Energy Strategy that the "single loudest message" they heard was the call for improved efficiency. That call, it is now clear, was ignored.

The outstanding feature of the Bush National Energy Strategy is its lack of new direction, its clear resolve to avoid any significant move toward an energy future based on efficient use of energy and renewable energy sources. Three key elements of the Bush Strategy illustrate the commitment to an energy economy based on the wasteful use of fossil fuels and nuclear power.

First, the provisions ostensibly aimed at energy efficiency are actually likely to increase energy use. For example, as Leon Lowery of Environmental Action explains, "Supposedly in the name of promoting competition in the utility business, Bush is proposing to free wholesale electric generators from the Public Utilities Holding Company Act. This would mean that any company that wanted to could build an electric generating plant, and sell electricity wholesale to utilities (but not to consumers, i.e. retail) and be outside the normal state utility regulatory process." Thus if Westinghouse, a nuclear-plant manufacturer, wanted to operate its own nuclear facility and sell the electricity it generated to a utility, it could. The Bush administration defines this as "efficient," because it will promote competition in the electricity generating field; it will do nothing to reduce overall energy used and improve energy efficiency, however. In fact, as Lowery notes, the Bush proposal "raises a host of consumer issues, including loss of accountability, as well as environmental issues because it undermines incentives to meet electricity demand with efficiency rather than new production. This plan is emphatically anti-efficiency. It would create a major new outlet for the urge to build and sell electricity."

"Unless oil *consumption* in the United States is reduced, the nation's reliance on oil from the Middle East . . . will continue."

Bush has completely ignored true energy efficiency strategies. Yet technologies are available that could more than double the energy efficiency of appliances, lighting, electric motors, space heating, electrical generation and industrial processes. Amory Lovins of the Rocky Mountain Institute estimates that U.S. electricity use could be cut in half at a savings of $50 billion per year without any reduction in standard of living. Due largely to improvements in energy efficiency, U.S. energy consumption has held virtually steady over the last 15 years even though the economy has grown by nearly 40 percent. "Bush's plan ignores this record, and the even greater potential offered by efficiency," says Lowery.

The Pursuit of Energy

The proposal's second key element is the pursuit of "energy security," the leading element of which is the opening of the Arctic National Wildlife Refuge for drilling by oil companies. In a plan released just weeks before the National Energy Strategy, Bush made clear his intention to spur offshore oil production as well. In addition, the administration plan would continue tax credits for the domestic oil and gas industry, and provide a variety of new forms of "regulatory relief" for those industries.

This approach ignores the inescapable fact that unless oil *consumption* in the United States is reduced, the nation's reliance on oil from the Middle East and other parts of the world will continue. Two-thirds of all the known oil reserves in the world lie in the Middle East. Only 4 percent lie in North America. According to Dorothy Smith, of the Greenpeace Ocean Ecology Campaign, "The Department of Interior's own analyses project that undeveloped U.S. offshore reserves and the Arctic National Wildlife Refuge combined hold no more than two years' worth of oil at current rates of consumption. But Bush seems to think that by sacrificing these areas to oil production, he can please his friends in the oil business and fool the public into thinking that he's got an energy strategy worthy of the name."

Drilling in the Arctic National Wildlife Refuge would have its own high environmental costs. The air pollution, toxic spillage and solid waste production normally associated with oil drilling could devastate the fragile arctic ecosystem. "Putting up large-scale oil developments there will ruin the refuge," says Larry Landry of the Northern Alaska Environmental Center. "Wildlife populations will very likely take it on the chin," he adds.

The third prong of the waste-more, supply-more strategy is a "revitalization" of nuclear power. The nuclear power plant licensing process and nuclear waste disposal are identified as key impediments to such a revitalization. Under the Bush plan, licensing would be made easier and speedier by cutting citizens out of the process. While nuclear power plants must currently go through a public permitting process for both a construction permit (before the plant is built) and an operating permit (after the plant is built but before start-up), Bush would grant the nuclear industry's long-held wish for "one-step" licensing—eliminating the post-construction licensing process. Streamlining the licensing process would limit options for anti-nuclear activists who have used it to expose nuclear plant deficiencies and safety hazards. "The one-step provision is blatantly anti-democratic," says Scott Denman, Executive Director of the Safe Energy Communications Council. "Decisions about energy production and use that directly affect the public already suffer from too little citizen participation—not too much."

"Bush's proposal on nuclear waste is equally dangerous," adds Denman. "He is proposing to unravel the 1982 Nuclear Waste Policy Act which barred the creation of new temporary waste storage sites until a permanent repository is found. With no permanent repository yet sited, the result of removing that prohibition will be to create new *de facto* permanent waste sites without the oversight and site-selection criteria that would otherwise be required."

No Alternatives

According to the National Energy Strategy, the United States does have at least one abundant renewable energy resource—garbage. While the administration ultimately rejected a temporary tax credit for producers of solar, wind, geothermal and bio-mass energy which had been included in all of the NES drafts, garbage burning will be researched and promoted under the plan. "It's ludicrous," says Stevenson of Greenpeace. "Burning garbage wastes energy by destroying resources that could be reused and recycled, and poisons the environment. Meanwhile, legitimate renewable resources are under-utilized and under-developed because they are trying to compete in a marketplace with heavily subsidized fossil-fuel and nuclear sources."

"Legitimate renewable resources are under-utilized and under-developed."

Stevenson contrasts the U.S. approach with that of other countries, which "have gotten the message that the energy future is in efficiency and renewables, and are promoting investment in that direction. If government and businesses in this country won't recognize that reality," she says, "we will be left behind by nations that do." Her point is illustrated by trends in the world market for photovoltaic cells (PV), a market which is doubling every three years. In 1980, the U.S. share of the world PV market was 75 percent. By 1988, it had fallen to 32 percent. Japan's share increased from 15 percent in 1980 to 37 percent in 1988.

Though two-thirds of the nation's petroleum consumption occurs in the transportation sector, the only alternative proposed by Bush in this area is to require automobile fleets to use alcohol fuels. The 175 million cars and trucks on U.S. roads use over 300 million gallons of gas each day. The energy wasted by cars and trucks is enormous. Cars in the United States get 20 miles to the gallon, on average; raising that average to 32 miles per gallon would save as much oil as the United States imports from the entire Middle East. Yet, Bush kept any increase in the automobile fuel efficiency requirements out of his energy plan. Support for transit and land-use planning that could actually reduce reliance on the automobile is wholly absent as well.

Bush's transportation plan, announced shortly

before the NES, would actually increase motor vehicle use. "We must not forget that oil is a reason why our valiant men and women find themselves fighting and dying in the Persian Gulf war," says Ruth Caplan, Executive Director of Environmental Action. "Yet the President can only speak of 'a new national highway system,' while maintaining his staunch opposition to higher fuel-efficiency standards for our cars. It's a recipe for continued oil gluttony."

Bush's disdain for real energy alternatives continues a pattern set when Ronald Reagan took office. Instead of maintaining the energy efficiency and renewable energy programs begun under Jimmy Carter, the Reagan and Bush administrations:

- Cut funding for research and development of solar and other renewable energy technologies at the federal Department of Energy by 85 percent between 1980 and 1990;
- Cut federal funding for mass transit by half from 1981 to 1989; and
- Granted automobile companies' petitions to weaken car fuel efficiency requirements in 1986, 1987 and 1989. In addition, Bush successfully fought for the defeat of a Senate bill in September 1990 that would have increased auto fuel-efficiency by more than 40 percent.

"Congress' track record of taking on the oil, auto, coal, utility, nuclear and highway lobbies is quite poor."

The political contours of the debate over the National Energy Strategy have already been set. When a near-final draft of the NES was leaked in early February 1991, oil industry executives quickly acknowledged their pleasure. Douglas G. Elmets, spokesperson for ARCO, told the *New York Times* that, "If this is as described, it certainly sounds very positive, not only for the industry, but for the consumer." Environmentalists blasted the plan, and some members of Congress, such as senators Albert Gore, D-TN, and Tim Wirth, D-CO, indicated their opposition.

But the Bush plan has set the terms of the debate, and the U.S. Congress' track record of taking on the oil, auto, coal, utility, nuclear and highway lobbies is quite poor. Congress' inability to stand up to corporate interests and do what is best for the United States and the world was illustrated again in 1990 when it watered down the Clean Air Act to the point where actually achieving clean air became a goal for decades hence, not a genuine political objective. While the Congress may well add a few efficiency measures, and blunt some of the worst proposals in the Bush plan, that will not be enough to produce what is really needed: a dramatic redirection of America's energy economy.

To Change the Course

Such a change will only come from a broad and powerful movement, broader and more powerful than the existing environmental movement. It will have to be active regionally and locally as well as nationally. It will have to include businesses that want to cut their energy bills through conservation; farmers that want to cut their costs and unhook from energy-intensive, chemical-intensive farming methods; local citizens who want to prevent their communities from being carved apart by new highways; community groups who are already active on toxics issues and who recognize that the flipside of petro-chemicals is petro-energy; and the many people who were forced by the Persian Gulf War to recognize that "security" will come neither from weapons nor from oil, but from a more self-reliant economy and an understanding of global interdependence. The task is as enormous as it is vital, but the world cannot afford a delay.

Fossil-Fuel Use Should Be Cut to Reduce the Greenhouse Effect

Natural Resources Defense Council

About the Author: *The Natural Resources Defense Council is an organization that addresses many environmental concerns, from the destruction of tropical rain forests to toxic wastes.*

The beginning of a new year and decade brings with it the unmet environmental challenges of the past and the hope and confidence that we can make amends for our assault on nature. At the top of our list of priorities for the nineties is reducing emissions of carbon dioxide [CO_2] and other pollutants—notably chlorofluorocarbons (CFCs), nitrous oxide, and methane—which are subtly altering the climate of the entire planet.

While it is difficult to imagine the worst consequences of the greenhouse effect, the growing consensus of atmospheric scientists is that the threat of global warming is serious and demands immediate action. In the United States alone, climatic disruptions could eliminate virtually all Eastern forests, undermine the productivity of the Midwestern grain belt, and destroy up to 80 percent of our coastal wetlands.

Effects in many other nations, whose economic and environmental systems are more fragile, would be even more devastating. The resulting stresses could fracture the economic, social, and political structures that protect the peace and, for the first time in all of human history, hold the promise of providing a full, dignified life for all people.

Natural Resources Defense Council, "Cooling the Greenhouse," *The Amicus Journal,* Winter 1990, © 1990, *The Amicus Journal,* a publication of the Natural Resources Defense Council. Reprinted with permission.

We have no time to waste. The June 1988 Toronto Conference on the Changing Atmosphere recommended that the world reduce its CO_2 emissions 20 percent by the year 2005, just fifteen years from now. The United States and other industrialized nations must take the lead. NRDC [Natural Resources Defense Council] recommends that the United States take a first step against global warming by reducing its CO_2 emissions at least 20 percent below 1987 levels by the year 2000.

To this end, the federal government should adopt an energy planning process that treats energy conservation as a resource on equal footing with energy production and considers all the economic and environmental costs associated with each supply and conservation option. This "least-cost" method can enable our nation to identify many other cost-effective conservation measures and guide other relevant federal decisions, such as those concerning energy research and development. A fee on fuels proportional to their carbon emissions would also provide an important market signal and reduce the budget deficit. Eventually, these actions would save energy consumers tens to hundreds of billions of dollars each year.

"Worldwide CO_2 emissions eventually must be reduced 50 percent or more to minimize climate change."

The federal government should phase out its massive program to demonstrate new methods of burning coal with less emissions of the pollutants causing acid rain. The so-called "Clean Coal" program would account for over 30 percent of the federal energy research and development budget, and its success would only increase the global warming problem. Similarly, nuclear power is not the answer to rapid CO_2 emissions. Lead times for building nuclear plants are too long, and public acceptance too low, for nuclear power to contribute to the year 2000 goal. We will need new energy supplies in the twenty-first century, but that energy should come from renewable sources.

Reducing U.S. emissions of CO_2 by 20 percent is vital, but the United States now accounts for only about 23 percent of global CO_2 emissions, and that proportion is dropping. Without major reductions in CO_2 emissions from other parts of the world, we cannot solve the problem. Worldwide CO_2 emissions eventually must be reduced 50 percent or more to minimize climate change.

At the outset, we must persuade the other Western industrialized nations to make equally rapid reductions in their CO_2 emissions. We must also take steps to ensure that we are joined as soon as possible by the Soviet Union and the Eastern European nations, which are extremely inefficient in their use of energy.

The major developing countries must join in efforts to minimize CO_2 emissions as soon as possible. Their emissions now account for only about 30 percent of the world total, but are likely—in view of fast-growing human populations and very low per capita energy use (the average Indian now uses about 4 percent of the energy consumed by the average citizen of the United States)—to account for some 50 percent or more by the year 2025. If these trends continue, CO_2 emissions from these countries could defeat the objective of climate stabilization.

The International Dimension

The United States should use scientific evidence, diplomacy, and offers of assistance to persuade developing nations to join the effort to minimize global warming. We must provide increased attention and assistance to stabilizing the world's population.

The United States should use its bilateral foreign aid program and its support of the World Bank and other multilateral actions with respect to energy. The same means should be used to stimulate actions to reduce tropical deforestation, now thought to account for 10 to 30 percent of global CO_2 emissions, and to promote tree planting. Aggressive forestry actions could reverse this situation and make the forests a net sink for CO_2.

We also must act against CFCs and their chemical relatives, which not only contribute to global warming but deplete the ozone layer. Even under the recently negotiated Montreal Protocol, the concentration of these chemicals in the atmosphere will double before 2050. They must be phased out as rapidly as possible. We believe that the United States should phase out CFCs domestically by 1995. . . .

"Some global warming appears to be irreversible."

NRDC also recommends a global treaty to establish a greenhouse gas reduction goal and allocate responsibilities among nations, including a commitment from wealthier countries to increased research into non-CO_2 energy supply technologies and aid to poorer countries working to meet the treaty requirements. Funds might be generated by a worldwide greenhouse gas emission fee or a levy on wealthy nations in proportion to their emissions. The United Nations Environment Program should be the forum for negotiation of the agreement.

Some global warming appears to be irreversible and inevitable due to past emissions of greenhouse gases. We must, therefore, begin now to take adaptive actions. All coastal planning and construction should take fully into account the likely impacts of global-warming-induced sea-level rise. Agricultural research should focus on alternative crops and farming methods that can withstand the likely changes. Forestry planning should aim to conserve large ecosystems that have the flexibility to adapt.

Global warming is a real, immediate threat to the habitability of our planet. We cannot make up for the damage we have already caused, but we can begin now to minimize the rate and extent of climate chaos.

Fossil-Fuel Use Should Be Cut to Reduce Air Pollution

James J. MacKenzie and Mohamed T. El-Ashry

About the Authors: *James J. MacKenzie is a senior associate in the Program in Climate, Energy, and Pollution at World Resources Institute (WRI) in Washington, D.C. Mohamed T. El-Ashry is the institute's vice president for research and policy affairs. WRI is a research and policy center that looks for ways to meet human needs and nurture economic growth without degrading the planet and its resources.*

Across the United States, forest trees and crops are under attack. Many ponderosa and Jeffrey pines in southern California's San Bernardino National Forest have died over the past three decades. On well over 100,000 acres of this land, pines are aging faster, growing slower, and succumbing more easily to insects than they did 30 years ago. Likewise, at elevations above 2,500 feet in the Northeast, half the red spruce trees that appeared healthy in the early 1960s are now dead. And throughout the Farm Belt, certain major crops are growing slower than they should be or have leaves that are misshapen or discolored.

What's the assailant? In each case the evidence points to air pollution. Together or separately, ozone and acid precipitation from the burning of fossil fuels are weakening trees in U.S. forests, leaving them more vulnerable to natural stresses. And ozone is adding to the burdens of the American farmer by reducing agricultural productivity to the tune of billions of dollars a year.

James J. MacKenzie and Mohamed T. El-Ashry, "Ill Winds: Air Pollution's Toll on Trees and Crops," *Technology Review*, April 1989. Reprinted with permission from *Technology Review*, © 1989.

Although scientists have suspected for some time that air pollution is contributing to forest declines—especially in central Europe, where the problem is most severe—evidence has until recently been scant. It is still hard to separate out the role of other factors, such as weather extremes, insects, and competition among species. But now that numerous studies of forest declines have been completed in both the United States and Europe, the evidence implicating airborne pollution is too strong to ignore. The case that air pollution, primarily ozone, is damaging crops is even more compelling. This is partly because greater resources have been devoted to crop research and partly because the ecology of agricultural systems is simpler.

As crops and trees are added to the list of entities threatened by air pollution—which already includes lakes and streams, aquatic life, materials, scenic views, and human health—the policy imperatives for the United States are now clearer than ever: the federal government must devise a more effective air-pollution control strategy and work with industry to develop clean sources of energy. . . .

"Ozone and acid precipitation from the burning of fossil fuels are weakening trees in U.S. forests."

Without an effective abatement strategy for the long term, the harm caused by ozone and acid precipitation will only get worse. The use of fossil fuels is expected to increase significantly in the coming decades. Unchecked, this growth will not only hasten direct damage to the environment but also aggravate the global greenhouse problem.

Besides the declines that have hit the San Bernardino National Forest and northeastern red spruce, there are other instances of dead and dying trees that cannot be explained by natural processes. . . .

At some high-elevation sites, trees have been declining for almost 30 years; at others, symptoms have developed only within the past decade. What all these sites have in common, however, is that they are subject to high levels of air pollution. Most sites have much higher than average concentrations of ozone, which forms when nitrogen oxides and hydrocarbons (both produced when fossil fuels are burned in power plants, industrial boilers, and vehicles) undergo photochemical reactions in the air. . . .

"The use of fossil fuels is expected to increase significantly in the coming decades."

Acid deposition is also high at many sites, especially in eastern mountains, where forests may be covered in acidic fogs and clouds for up to 3,000 hours each year. The acids in question—nitric acid and sulfuric acid—form in the atmosphere from emissions of nitrogen oxides and another by-product of combustion, sulfur dioxide. They are carried to earth not only by clouds, fogs, and rain but also by dry deposition, in which gases and acid particles settle out of the air. . . .

Acid Precipitation

Of course, it's one thing to note that ozone and acid precipitation are present where trees are dying or being injured, but quite another to prove that these pollutants are contributing to the damage. In many cases the immediate cause of the decline is a natural stress such as winter cold, insects, or disease. In other cases, such as the declines in the San Bernardino mountains, the symptoms do not match those of any natural affliction. The present understanding is that these declines result from multiple stresses, with air pollution playing an important—and sometimes essential—role.

Various studies have shown that the levels of pollution at the affected sites can indeed weaken trees, causing nutrient imbalances and lowering the rate of photosynthesis. In the early 1950s, when damaged needles first appeared on San Bernardino's ponderosa pine, scientists enclosed some of the affected branches in chambers. They then fumigated the branches with the surrounding ozone-rich air, with filtered air, or with filtered air plus measured amounts of ozone. The branches in the filtered air improved, while the others continued to deteriorate. These and later fumigation experiments confirmed that ozone was causing the damage. . . .

Investigators have also identified a number of mechanisms by which acid deposition directly and indirectly injures trees. . . .

More threatening than direct damage to foliage are the many indirect changes that acids can bring about by altering soils. First, acid deposition can leach important nutrients out of the soil, replacing them with hydrogen ions and acidifying the soil in the process. Trees growing on such soil may suffer nutrient imbalances that lead to decline.

Second, high levels of acid deposition can release aluminum from minerals in the soil. Aluminum ions damage the fine roots of trees and block the uptake of calcium and magnesium. . . . Finally, because of its nitric-acid component, acid precipitation can overload forest ecosystems with nitrogen. By one estimate, seven times as much nitrogen is deposited at a high-elevation New Hampshire site as at low elevations. Nitrogen is a vital fertilizer, but if the soil is lacking in other nutrients, the growth spurred by an overdose of nitrogen can lead to nutrient imbalances. Too much nitrogen can also make a tree more susceptible to freezing or drying out in winter.

Of course, not all these mechanisms will apply at any one site. Moreover, the actual rate at which leaching occurs—and the amount of damage it causes—depends on the condition of the soil, the amount of acid that is deposited, and other factors.

So far, the declines have hit only those forests with very high pollution levels. Nevertheless, it's possible that the damage could eventually spread to less polluted areas. If it turns out that lower levels of acid deposition lead to significant nutrient leaching, the same injuries could result over a long period as result from higher levels over a short period. Thus, it may be only a matter of time before forest declines in the United States become as serious as those in West Germany, where trees of all important species at all elevations are showing signs of damage. . . .

Injured Crops

It's hard to put a dollar value on the trees that are now declining, since most are on state or federal land. But the economic losses that air pollution causes to crops have been estimated as part of the Environmental Protection Agency's National Crop Loss Assessment Network (NCLAN), a program that ended in 1987. According to Walter Heck, of the Department of Agriculture's Air Quality Research Program, who chaired NCLAN's research committee, cutting ozone levels by 50 percent would increase yields for four major crops—soybeans, corn, wheat, and peanuts—by up to $5 billion. Overall, ozone from burning fuel is estimated to reduce crop yields by 5 to 10 percent.

"Many of the steps we take to improve the air for crops and forests will reduce ill effects on human health and global climate."

Unlike forests, crops appear to suffer little or no damage from existing levels of acid deposition. Agricultural soils are well fertilized, so crops can readily replace the nutrients they may lose through leaching. But ozone is not so easily countered. . . .

In contrast to forest declines, which so far have been confined to scattered sites, the losses in crop productivity are pervasive. Various economic analyses estimate that in each of 13 states the losses exceed $100 million a year.

Turning the Tide

Air pollution is already taking a heavy toll on America's forests and crops. And the toll will rise unless more is done to curb emissions. Under 1989 regulations, nitrogen-oxide pollution from power plants is expected to double by 2030. Over the same period, nitrogen oxides from transportation, industry, residential and commercial heating, and waste incineration are likely to increase by 30 percent, and hydrocarbon emissions by 25 percent. By 2010, emissions of sulfur dioxide are expected to be no lower, and perhaps slightly higher, than they are today.

Fortunately, there is much that we as a nation can do to turn the tide. And many of the steps we take to improve the air for crops and forests will reduce ill effects on human health and global climate. In lessening the threat to one domain, we will also lessen the threat to others.

A good place to start is with more effective legislation for reducing emissions of sulfur dioxide and nitrogen oxides. . . .

New legislation should impose state or regional caps on the release of sulfur and nitrogen. To decrease sulfur-dioxide pollution (80 percent of which comes from power plants and industrial processes), the act should require a 50 percent reduction in such emissions—to 10 million tons a year—over a period of, say, 10 years. This goal is consistent with the National Academy of Sciences' conclusion that cutting acid deposition in half would probably protect sensitive aquatic life. It would also go a long way toward safeguarding trees and crops.

Each state should then be permitted to choose among various options for meeting the new goal as cost-effectively as possible. A state could encourage its coal-burning power plants to convert to "clean-coal" technologies (such as fluidized-bed combustion) or to switch from one fuel to another—say, from coal to natural gas.

Or a state could trade emissions, both among sources within its borders and with neighboring states. For example, if a power plant in one state could convert to clean coal more cheaply than a nearby plant in a second state, the first might offer to convert its plant in exchange for some concession from the second.

A state might also encourage the use of more efficient appliances, lighting, and other equipment, to reduce the need for electricity. . . .

Cutting Emissions

Nitrogen-oxide emissions (of which the main source is motor vehicles) can be reduced in much the same way. A reasonable goal would be to cut annual emissions by 5 million tons over the next decade—a 25-percent reduction from 1986 levels but a much larger reduction from projected growth. This goal can be reached through a combination of stricter emission limits for cars, buses, and trucks; strengthened inspection and maintenance; and stronger measures to prevent tampering with pollution-control equipment. The use of cleaner fuels, such as compressed natural gas, in commercial fleets and urban buses would also be beneficial.

"The United States should be preparing itself for the inevitable shift to nonfossil energy sources."

A number of other straightforward measures could reduce pollution by cutting the total number of vehicle miles traveled. For example, cities could encourage greater use of public transit, reserve parking spaces for car pools, remove subsidies for other spaces, and designate more traffic lanes for car pools and bicycles.

To raise fuel efficiency, and thereby lower pollution levels, federal and state governments could increase fuel taxes, as well as tax new vehicles according to their gas mileage. (Annual registration fees could also vary this way.) In addition, governments might buy mostly ultra-efficient vehicles for their own fleets.

These and other measures for cutting emissions would lessen the damage to trees and crops and reduce threats to public health. Yet two serious fuel-related problems would remain: growing U.S. reliance on foreign oil and the steady increase in global warming. In principle, the nation could solve the first problem by converting to coal, which is abundant and which could be turned into methanol (to replace gasoline) or synthetic crude oil. But this would aggravate the second problem. The burning of fossil fuels, especially coal, releases vast amounts of carbon dioxide that contribute to greenhouse warming. The only way to protect natural resources *and* contain oil imports *and* slow global climate change is to burn much less fossil fuel.

Thus, at the same time it is working to curb emissions, the United States should be preparing itself for the inevitable shift to nonfossil energy sources. The renewable-energy technologies—solar cells, wind turbines, hydropower, geothermal energy—are strong candidates to assume the burden of future power production. "Second-generation" nuclear technologies (small, inherently safe fission reactors) offer another, though less certain, option. In the long term, transportation is also due for some profound changes, perhaps a switch to electric or hydrogen-powered cars, with the electricity or hydrogen ultimately derived from nonfossil sources.

These technologies will require considerable R&D [research and development], and probably major government initiatives, before they can be applied on a grand scale. But once they are in place, nonfossil technologies will alleviate a whole range of national problems—not just tree and crop damage but urban air pollution, climate change, and dependency on foreign oil as well.

Alternatives to Fossil Fuels Are Necessary

Christopher Flavin and Nicholas Lenssen

About the Authors: *Christopher Flavin is vice president of the Worldwatch Institute, a research center that focuses on environmental issues. Nicholas Lenssen is a research associate with Worldwatch.*

The world is now lurching from one energy crisis to another, threatening at every turn to derail the global economy or disrupt its environmental support systems. The nineties are likely to be plagued by more frequent and more severe energy crises than ever before.

While the failure of societies to redirect their energy futures is in one sense a failure of policy, it is also a failure of vision. Political leaders have little concept of an energy system not based on fossil fuels. Nor do they seem to see that an alternative approach is possible. Society is entering a period of inevitable and rapid change in its energy systems, with little idea of where we are headed or how our course can be shaped.

The World Energy Conference, a triennial gathering of energy officials and experts, concluded in late 1989 that by 2020 the world would be using 75 percent more energy, and that most of it would be supplied by coal, oil, and nuclear power. Yet this business-as-usual scenario would not lead to a more pleasant version of the status quo—the apparent goal of many planners.

Such an approach would eventually entail relying on the Persian Gulf for more than two thirds of the world's oil, compared with 26 percent today. It would involve building three times as many nuclear plants in the next 30 years as in the past 30, accompanied by more frequent nuclear accidents and growing stockpiles of nuclear waste and plutonium. It would accelerate global warming as carbon dioxide emissions soared above today's levels. And the increasing centralization and growing scale of energy systems would require tight police supervision and restrictions on public participation. This picture of the future is neither attractive nor ultimately plausible.

Three major considerations are now forcing the world's energy systems in a different direction. The first is the availability of fossil fuels, particularly the most economical and versatile one—petroleum. The constraint is not the global resource base but the geographical and political limits of having nearly two thirds of the world's current oil reserves in the Persian Gulf region.

"The world is now lurching from one energy crisis to another, threatening at every turn to derail the global economy."

The second limit is environmental—the capacity of the world to cope with the overwhelming burden of pollution that is emitted by a $20-trillion world economy run on fossil fuels. The most intractable load is the nearly 6 billion tons of carbon added to the atmosphere each year. As no technical fix appears likely for this problem, slowing global warming will mean placing limits on fossil fuel combustion.

The third constraint is social and political. In recent years, citizens around the globe have rebelled against the energy "solutions" their governments pursue. The people of West Germany, for example, effectively ruled out nuclear expansion during the eighties, and those in the Soviet Republics are in the process of doing the same. Less controversial technologies have also been stopped. Coal-fired power plants are rarely con-

Christopher Flavin and Nicholas Lenssen, "Designing a Sustainable Energy System," *State of the World 1991*. Reprinted with permission of the Worldwatch Institute.

structed now in the northeastern United States, and the Indian government's efforts to build new hydro dams have met with massive public protest. Political leaders around the world are beginning to realize that people's passionate concerns cannot be swept aside by a tide of technocratic policymaking.

Powerful economic, environmental, and social forces are now pushing the world toward a very different energy system in the decades ahead. But what might it look like? Ultimately, a sustainable economy must operate with much lower levels of fossil fuels, and probably without nuclear power. It would likely derive its power from solar resources replenished daily by incoming sunlight and by geothermal energy—resources available in far greater abundance than fossil fuels. It would also need to be much more energy-efficient, since renewable energy is unlikely ever to be as cheap as oil has been.

A solar economy will involve the creation of whole new industries and the restructuring of the job market. Ultimately, new transportation systems will likely evolve, and both cities and agriculture will be changed. The challenge ahead is in part one of continuing to develop new technologies that use energy efficiently and harness renewable resources economically. But the most important challenge is political: societies need to overcome narrow economic interests and revamp energy policies in order to develop energy systems that future generations can count on.

Middle Eastern Oil

Just four decades ago, world oil consumption was one sixth the current level, with half of it being used in North America alone. As recently as 20 years ago, the petroleum economy had touched the lives of only a tiny fraction of humanity. While oil dependence may seem to our generation to be inevitable and permanent, it could turn out to be even more fleeting than the 200-year age of coal that preceded it.

In the immediate future, a chaotic world oil market may do the most to knock world energy trends off their current course. When Iraq's tanks rumbled into Kuwait in August 1990, the world faced the third oil shock in the space of just 17 years. This invasion, which immediately raised Iraq's share of world oil reserves from 10 to nearly 20 percent, caused a 170-percent increase in oil prices in three months and led to near panic in world financial markets.

"A sustainable economy must operate with much lower levels of fossil fuels."

The forerunners to this crisis were the failed energy policies that allowed oil-consuming nations—both industrial and developing—to increase greatly their dependence on Middle Eastern oil in the late eighties. The world's addiction to cheap oil is as destructive and hard to break as an alcoholic's need for a drink. Since 1986, when oil prices fell back below $20 per barrel, the move toward more efficient homes, cars, and factories that began in the mid-seventies slowed to a crawl. As a result, world oil demand shot up by almost 5 million barrels per day, or nearly 10 percent.

Virtually all the extra oil now being consumed is supplied by a handful of countries in the Middle East, a region that faces the stresses of rapidly growing populations, autocratic political systems, rampant poverty, and a deadly arms race. Not only is the world addicted to cheap oil, but the largest liquor store is in a very dangerous neighborhood.

The uneven distribution of world petroleum resources is growing more lopsided all the time. While the Persian Gulf region had 55 percent of proven global reserves as recently as 1980, by 1989 that figure reached 65 percent. Most of the nations in that area have at least 100 years of proven reserves left at current extraction rates, compared with less than 20 years' worth in Europe, North America, and the Soviet Union.

Outside the Middle East, much of the cheap

oil has already been consumed. In the Soviet Union and the United States—still the world's leading producers—output is now declining. The U.S. fall is hardly surprising since the country's heavily exploited oil fields have only 4 percent of global reserves while still accounting for 12 percent of production. Whereas the average oil well in Saudi Arabia produces 9,000 barrels per day, the average well in the United States produces 15. The Soviet Union also appears poised for a steep decline as it cuts its disproportionate investments in the petroleum sector. Infusions of western oil technology could slow the decline but are unlikely to stop it entirely.

If the past is any guide—and there is every reason to think it is—the nations of the Persian Gulf are in no position to provide a steady long-term supply of oil. To rely increasingly on them would set the world up for an unending series of price run-ups, economic crises, and oil wars. Developing countries with large debt burdens are particularly vulnerable to a continuation of the oil-price roller coaster. India, for example, was forced to cut its oil consumption by a remarkable 25 percent as prices skyrocketed in late 1990.

Reducing Petroleum Dependence

Oil-consuming nations therefore face the imperative of reducing their petroleum dependence. But by how much? Current oil use per person averages 4.5 barrels a year, but it ranges from 24 barrels in the United States to 12 in Western Europe and less than 1 in sub-Saharan Africa. To stretch out oil supplies and reduce the environmental impacts of its use, the world is unlikely to be able to burn more than about 30 million barrels a day by the year 2030—one half the current level. Given population increases, this would allow for an average of just 1.2 barrels per person worldwide by then, implying extensive changes in the global energy economy.

The capacity of the global biosphere to absorb the emissions of a fossil-fuel-based energy system may in the end prove even more constraining than the limits posed by oil. Nearly 6 billion tons of carbon are spewed into the air each year in the form of carbon dioxide, a greenhouse gas that is building steadily in the atmosphere and gradually heating the planet. Although these concentrations rise slowly, future climate disruptions are likely to be abrupt and catastrophic.

"The world's addiction to cheap oil is as destructive and hard to break as an alcoholic's need for a drink."

Despite the vast public attention paid to the problem of global warming over the past three years, the amount of carbon being released annually has risen by 400 million tons since 1986—exactly the opposite of what many scientists believe is necessary.

A major scientific study released in 1990 by the United Nations-commissioned Intergovernmental Panel on Climate Change confirmed that a rapid and highly disruptive increase in global temperatures would occur unless emissions are cut. Upon releasing the report, Dr. John Houghton, head of the British Meteorological Service, noted that it represented "remarkable consensus," with fewer than 10 of 200 scientists dissenting. Although major cuts in carbon dioxide emissions will take decades to accomplish, the targets will be far more difficult to achieve if emissions continue to rise.

Somewhat balancing this bleak global trend, 15 nations have produced plans to limit their production of carbon dioxide—and, by implication, their use of fossil fuels. The leader is Germany, which aims to cut emissions within the former West Germany by 25 percent over the next 15 years. Yet in order to stabilize the atmospheric concentration of carbon dioxide, scientists believe global emissions must eventually be cut by at least 60-80 percent—to about 2 billion tons annually.

A world that produces only that much carbon

a year will be far different from one that produces 6 billion. Per capita carbon emissions some 40 years from now would need to be one fourth the level in Western Europe today, given the inevitable growth in world population over the next few decades. These are stringent targets, especially when one considers that fossil fuels now account for 75 percent of world energy supplies.

An annual carbon budget of 2 billion tons can only be met if use of coal, the most carbon-intensive fossil fuel, is cut by roughly 90 percent. Small amounts of coal would still be burned in countries like China and India, which have large populations and only limited reserves of other fossil fuels. Oil shale and other synthetic fuels can be ruled out entirely due to their high carbon content.

For most nations, however, natural gas will likely be the predominant fossil fuel still used—as it produces roughly twice as much energy per kilogram of carbon released as coal does. Natural gas resources are also better disbursed than those of oil; the largest known reserves are, however, in the Middle East and the Soviet Union. Forty years from now, natural gas could still be producing as much energy as it does today.

"The nations of the Persian Gulf are in no position to provide a steady long-term supply of oil."

In a world with an energy system that is truly sustainable—economically and socially—nuclear power will probably not be a major source of energy. During the past 10 years, the pace of nuclear expansion has slowed almost to a halt in countries around the world. All existing reactors are scheduled to be retired within the next 40 years, and it seems likely that most will not be replaced.

Although there is much that can be debated about the new energy technologies to be deployed and the ways in which cities and economies will be restructured, one point is clear: a sustainable energy system is possible only if energy efficiency is vastly improved. Overall, the world will have to be producing goods and services with a third to half as much energy as today. The member countries of the Organisation for Economic Co-operation and Development have lowered their energy use per unit of gross national product 24 percent since 1973, but plenty of opportunities for further improvement remain. The Soviet Union, Eastern Europe, and developing countries have an even larger, untapped potential. . . .

Quadrupling the output of renewable energy is also essential to achieving a sustainable energy system in the foreseeable future. This will entail expanding the use of biomass and hydropower, but more importantly will require that solar and geothermal energy become a major part of the world energy mix.

Energy Transition

The technologies are at hand to initiate this historic energy transition, but it will not occur without major changes in energy policy. The first step is to redirect a host of government policies so that they are aimed at achieving the central goals of improving energy efficiency and reducing fossil fuel use. The purchase of efficient cars could be rewarded, for example, and gas guzzlers discouraged through government levies. In areas such as requiring tighter building standards or encouraging local transportation alternatives to the automobile, state and local governments can play a leading role. . . .

Last, but not least, governments will need to raise taxes on fossil fuels so that prices reflect the full security and environmental costs involved. This would provide a great boost to the development of energy-efficient and renewable energy technologies.

The U.S. Should Decrease Its Dependence on Foreign Oil

Susan Dentzer

About the Author: *Susan Dentzer is a senior writer for* U.S. News & World Report.

Experts say the nation needs a multipronged plan to cut its foreign oil intake: A renewed emphasis on conservation, a push to produce more oil and gas and long-range development of alternative energy sources. Prompted by a range of threats—from the Gulf crisis to fears of global warming—the pendulum may be swinging back from Ronald Reagan's laissez-faire policies, which gutted federal outlays to develop new energy supplies and mainly left matters to the market. But despite large strides in conservation in some areas—industry's use of oil, for example, has dropped 10 percent—the U.S. is now twice as dependent on Persian Gulf oil as in 1973. A new moral equivalent of war on energy is looking more attractive. . . .

At the same time, the stage is set for fierce debate over precisely what form an energy plan should take. As Congress and the administration seek to forge the nation's response, "We're going to see a lot of agreement on the need for an energy policy, and darn little agreement on what it will be," says James Gustave Speth, president of the Washington-based World Resources Institute. Battles loom over thorny issues like offshore drilling as oil producers press to exploit America's few remaining petroleum frontiers. Pro-conservation forces will fight car manufacturers for higher auto fuel-efficiency standards, and a showdown looms between Congress and the Bush administration over increasing the size of the U.S. Strategic Petroleum Reserve. In the end, the nation may turn out to have less an energy problem than a tough political one. For it is doubtful that America is willing to pay the steep price necessary to lessen its dependence on foreign oil, now equal to about half of U.S. oil consumption.

"Stanching further energy losses could require making cars more fuel efficient and battling Americans' born-again love of gas guzzlers."

Aside from the possible hike in the gasoline tax—which might be part of a deal to cut the federal budget deficit—little immediate action is likely. . . . Making changes in the giant $250 billion U.S. energy sector requires long lead times, since new oil-drilling rigs and thousand-mile gas pipelines can't be created overnight.

The wisdom of many policies will also depend largely on the future price of oil. . . . Given budget pressures, higher federal spending on research for energy alternatives will be hard to justify if oil falls back to $18. On the other hand, a price above $25 a barrel "does more than anything we can do legislatively" to foster conservation, says Jack Riggs, staff director for the House Subcommittee on Energy and Power.

Despite the obstacles, the U.S. could make rapid headway by plugging holes in its leakiest energy-using sectors. Factories, homes and appliances could all be made more energy-efficient, but the biggest candidate for improvement is transportation, which accounts for almost two thirds of all U.S. oil consumption and which now uses 20 percent more oil than it did in 1973. Jet-fuel consumption has skyrocketed, rising 28 percent over the past five years to 1.3 million barrels a day, notes John Lichtblau, chair-

Susan Dentzer, "What to Do About Oil?" *U.S. News & World Report*, September 10, 1990.

man of the Petroleum Industry Research Foundation. Meanwhile, America's 184 million cars, trucks, buses and other vehicles now swallow half the 17 million barrels of oil that the nation consumes daily.

Much of the impetus behind this growth has been cheap gasoline, priced lower in real terms until recently than at any other time in the past 50 years. One key to lowering vehicle-fuel consumption is thus raising the price of gas. The Gulf conflict may have accomplished that for now, but another option is to hike federal motor-fuel vehicle taxes, now at 9.1 cents per gallon of gas and essentially unchanged since the early 1980s. An increase of 12 cents a gallon would raise about $12 billion a year in new revenues, bringing the gas taxes a bit closer to the levels of industrialized nations such as Japan, where the tax is equivalent to $1.62 a gallon. More important, a tax hike would function as a price mechanism to spur consumers to conserve—taking more mass transit or even following the DOE's [Department of Energy] advice to put enough air in their tires.

"A national energy policy will have to balance a variety of interests if it is to gain widespread support."

Stanching further energy losses could require making cars more fuel efficient and battling Americans' born-again love of gas guzzlers. Thanks to lighter materials and massive re-engineering the fuel efficiency of the average 1989-model car was almost double that of the average 1974 model: 28.3 miles per gallon vs. 14.2. But "by far the biggest source of oil available to the U.S. is the potential for higher gas mileage in our motor vehicles," contends Bill Magavern of the U.S. Public Interest Research Group. Senator Richard Bryan, Democrat of Nevada, has proposed raising the average fuel efficiency for new cars to 40 miles per gallon by the year 2000; proponents claim the measure would lower oil consumption by 2.8 million barrels a day. Yet automakers have long argued that consumers would balk at paying sharply higher prices for re-engineered cars. One alternative might be slapping special taxes on cars with low fuel efficiency, using the proceeds to pay rebates to shoppers who purchased cars that ran at least 40 miles to the gallon.

Search for New Oil Reserves

Roughly 75 percent of the world's proven oil reserves lie in OPEC [Organization of Petroleum Exporting Countries] nations, while the U.S., the most heavily drilled country in the world, has only 4 percent. And America's aging oil fields are slowly petering out; domestic U.S. oil production has declined from a high of about 9.7 million barrels a day in 1970 to about 7.7 million barrels today. Even production on Alaska's North Slope near famed Prudhoe Bay—the boon that helped rescue America from its last energy crisis—is waning. Oil producers insist that the United States now has no choice but to launch exploratory drilling in places thought to harbor oil. At the top of their list are areas off the coasts of California, Washington and Oregon, and in the Alaska National Wildlife Refuge, or ANWR (pronounced "Anwar"), about 60 miles east of Prudhoe Bay where an estimated 3.2 billion barrels of oil might be found.

The thirst for reserves could now prompt the biggest clash in years between Big Oil's backers and the growing environmental movement, which opposes what it views as an effort to "Drain America First.". . . "Our position is that we can [drill in ANWR] in an environmentally sound manner," contends Charles Davidson of Atlantic Richfield's ARCO Alaska division. "You don't turn a national sanctuary for wildlife into an industrial park," counters Lisa Speer of the Natural Resources Defense Council.

Even if new reserves are found, they'll be no quick fix: It could take up to a decade to develop production at ANWR. In the meantime, the administration will push for expanded tax incentives to encourage independent oil drillers, as well as settlement of disputes impeding planned offshore production. Chevron Corporation and state and local officials in California have locked horns over the company's bid to transfer oil by tanker—instead of by more costly and environmentally safe pipeline—from the new Point Arguello oil field off the coast of Santa Barbara. DOE Secretary James Watkins has sent emissaries to resolve the dispute, arguing that production from Point Arguello could increase U.S. oil supplies by up to 75,000 barrels a day.

Gasoline can be made from coal and oil squeezed from shale, but oil's price would have to top $40 to $60 a barrel to make the effort worthwhile. A better alternative may be to expand use of natural gas, a relatively clean fuel that can be substituted for oil in both transportation and stationary use. Americans will use 18.3 trillion cubic feet of natural gas in 1990, 93 percent of it from the U.S. and most of the rest from Canada. "There is plenty of natural gas available in the U.S. at reasonable prices," says Bill Hederman, senior vice president of ICF, Inc. . . .

Alternatives to Gasoline

Running cars on liquid-alcohol fuels such as methanol, made from natural gas, or ethanol, made from corn, is another alternative—though widespread use could quickly gobble up the nation's corn crop and all domestic supplies of natural gas. Using electricity to fuel cars might make more sense, since vehicle batteries could be charged at night when power plants have surplus capacity. General Motors and Chrysler are at work on electric cars and vans with a range of about 120 miles, making them contenders for many commercial uses and commuting to work. . . . Electric cars could be on the market by the mid-1990s, though expensive batteries could make them pricey.

But greater use of electricity would be no panacea. The nation already uses 50 percent more electricity than it did in 1973—and with electric use growing rapidly, areas along the Eastern seaboard and Florida are now contending with power shortages. To meet the growing need, the nation's electric power grid will have to be enlarged by one third to one half by the year 2000, at a cost of as much as $800 billion.

"Greater use of electricity would be no panacea."

These giant new power plants will have to be fired by coal or nuclear power—and because the clean-air amendments will make it harder and more expensive to burn coal, the atom may soon have the competitive edge. . . . The Nuclear Regulatory Commission is now seeking ways to cut the time it takes to license U.S. nuclear power plants, in part because related public hearings and lawsuits now stretch out plant development to as much as 11 years. Yet public fears of accidents and worries about nuclear-waste disposal seem certain to hem in new nukes. . . .

Wind-generated and solar energy seemed promising in the 1970s, but many people in these "renewable" energy industries "spent much of the 1980s sending out their résumés," notes energy specialist Daniel Yergin. A key reason was a 90 percent cut in DOE funds for renewable-energy research from fiscal 1980 to 1989. In a gesture to environmentalists, the White House is considering extending tax credits on investments in renewable energy.

That tactic underscores a key notion: A national energy policy will have to balance a variety of interests if it is to gain widespread support. "There are no miracle solutions" to America's energy problems, notes Yergin; moreover, it's a safe bet the nation could never fulfill all of its own energy needs. But like all addicts, the U.S. will have to tackle its oil dependence one day—and a few million barrels—at a time.

Should the U.S. Decrease Its Use of Fossil Fuels?

No: Fossil-Fuel Use Should Not Be Decreased

The U.S. Should Increase Domestic Oil Production

Charles J. DiBona

About the Author: *Charles J. DiBona is president of the American Petroleum Institute, a trade group which represents the U.S. petroleum industry.*

In February 1991, the Bush Administration released its long-awaited National Energy Strategy, laying out a range of options for federal energy policy that could reduce the nation's dependence on imported oil, especially from unstable sources like the Mideast. The proposal contains programs designed to increase the nation's domestic energy production, to improve the efficiency of energy consumption, and to encourage the use of alternative fuels.

Yet today these options are being hotly debated—and many policymakers object to proposals that would increase domestic energy production, in particular the exploration and development of a small portion of the Arctic National Wildlife Refuge (ANWR) and those portions of the outer continental shelf that were not closed off to oil and natural gas leasing by the Administration in July 1990. Instead, they are proposing that more aggressive conservation measures and the greater use of alternative forms of energy be used as substitutes for the domestic production of oil.

In fact, however, many of these proposals are unrealistic in economic terms—they do not take into consideration the real costs of conservation and alternative fuel measures. Before the nation adopts any energy policy, it must ask if it is possible for the United States to change its consumption and investment patterns so it will be able to halt growth in—or reduce—energy consumption while maintaining a reasonable rate of economic growth and a broad array of choices for its citizens. Similarly, it must realistically assess the economic and technological viability of fuel switching.

Energy policies which mandate changes in America's energy habits will necessarily bump up against fundamental economic realities: first, that the nation depends on the output of its industries and cannot quickly or easily change the energy-using capital equipment that runs them. Second, that it relies on a complex transportation network for commercial and private vehicles. And, third, that it cannot quickly or easily change the tens of millions of private investment decisions ordinary citizens have chosen to make on the location and size of their homes, the number and kind of automobiles they own, and how they get to work.

While cost-effective conservation is an important part of any energy strategy, aggressive conservation measures that seek to reduce energy consumption without regard to the burden they impose on the economy and consumers can be counterproductive.

"Policies which mandate changes in America's energy habits will necessarily bump up against . . . economic realities."

API [American Petroleum Institute] has closely studied the cause-and-effect relationship between energy demand and economic performance, and we have found that maintaining constant energy use with a growing economy would be painful to Americans and, as a practical matter, difficult to achieve. Just to hold U.S. energy consumption constant at 1987 levels through the year 2000, the overall price of energy—the weighted average of the prices for oil, natural gas, and coal—would have to rise, in real

Charles J. DiBona, "Balancing the National Energy Strategy," *Maritime Reporter/Engineering News,* April 1991. Reprinted with permission.

terms, to three to four times its current level. That means that real energy prices (in 1990 dollars) would have to rise, at least within the United States, to as much as $55 per barrel in oil equivalent terms over the period.

To reduce aggregate U.S. energy consumption by 10 percent in the same time frame, the real weighted average energy price would have to rise to five or six times present levels, or to as much as $87 per barrel. Of course, these dollar amounts represent the average price of all energy—the actual price of oil would be even higher.

"The conservation goals being proposed would require enormous sacrifice on the part of individual Americans."

These higher energy prices could dramatically slow economic growth and therefore reduce the disposable income of all Americans. Even small differences in economic growth rates, if they persist, result in large differences in absolute family income over the long term.

If the nation were to reduce energy use by as much as some have urged through government-imposed limits, normal rates of economic growth could be maintained only if the "saved" energy was replaced through an increase in the proportion of GNP [gross national product] going for investment.

This increase in investment would have to be on the order of 50 percent—that's approximately $500 billion of aggregate investment—and consumption of goods and services would have to be reduced by about 16 percent.

From this, it is clear that the conservation goals being proposed would require enormous sacrifice on the part of individual Americans, if economic growth is also to continue. If instead we accepted slower economic growth rates, there would be fewer jobs and a reduced standard of living.

Similarly, alternative fuels must play a key role in any balanced energy policy—but to be effective they must be affordable and technologically viable. Proposals that simply mandate the widespread introduction and use of alternative fuels ignore the technical and economic aspects of introducing these fuels, including the costs of changing the infrastructure of the U.S. transportation system.

Many oil companies have a large stake in alternative fuels research and development, including solar energy, hydrogen, shale, coal gasification, geothermal energy, and new uses for electricity as a source of transportation energy. Yet all these alternatives are currently constrained by economic, environmental, or technological limitations—and, although they have been on the market for quite a few years, their combined share of our energy market is still less than 1 percent. If the United States is to maintain a healthy and growing economy, it must not force uneconomic alternative fuels into use. Achieving our nation's goals for economic growth requires that alternative fuels satisfy cost-effectiveness criteria.

Clearly, both uneconomic conservation measures and the widespread mandating of expensive alternative fuels would have a punitive effect on the U.S. economy and on the lives of American citizens. It is clear that if the United States seriously wishes to reduce its dependence on insecure sources of imported oil and at the same time maintain a healthy economy, it must take steps to increase opportunities for domestic oil and gas production. Unfortunately, recent trends have gone the other way, and domestic oil production has continued to fall—by more than 25 percent in the past 10 years—to 7.2 million barrels a day.

A Good Start

The Administration's proposals are a start towards turning this trend around. It has proposed that the coastal plain of Alaska's Arctic National Wildlife Refuge be opened to oil and gas leasing. It has proposed responsible exploration and development of portions of the outer continental shelf—though not those most promising offshore areas of California, Florida, and other states that were closed in July 1990. It has proposed leasing of Naval Petroleum Reserves and oil pipeline deregulation. And it has proposed options that would increase natural gas use.

"Developing ANWR coastal plain oil could . . . increase employment nationwide by about 735,000 jobs."

Over the next 10 years, for example, untapped domestic reserves of oil and natural gas on government lands could add the equivalent of from 2 million to more than 4 million barrels of oil a day to domestic production. That is a significant amount. During the last week of January 1991, for comparison, U.S. oil imports totaled 6.6 million barrels a day.

At the same time, development of these reserves will help strengthen the economy. In ANWR alone, the amount of oil that potentially could be found will make the costs of exploration and development economically feasible. Moreover, according to a recent economic analysis prepared by Wharton Econometrics Forecasting Associates, developing ANWR coastal plain oil could boost the gross national product by $50.4 billion and increase employment nationwide by about 735,000 jobs by the year 2005. That study predicted that ANWR oil development would "stimulate U.S. investment, moderately temper the growth in world oil prices and significantly reduce U.S. petroleum imports and improve the U.S. trade balance."

Opening the ANWR coastal plain to environmentally sound exploration and development is clearly one of the Administration's most important energy initiatives, as it can serve not only to reduce the nation's dependence on imported oil, but also to bolster the nation's economy. At the same time, we hope and expect that this proposal could open debate on a similarly positive approach to development of parts to the nation's outer continental shelf that have been placed off limits by government leasing moratoria.

A Sound National Energy Strategy

These are goals worthy of a national energy strategy. We can only hope that as Congress debates and evaluates the Administration's options, it will reach a consensus on positive, realistic steps that will contribute to the nation's energy security and economic well being.

The U.S. Should Increase Its Use of Natural Gas

Robert Hefner III

About the Author: *Robert Hefner III is chairman of the GHK Company, an Oklahoma City-based natural gas exploration and production company.*

It's time to get serious about America's best energy source: natural gas. Natural gas is even in oversupply today—as evidenced by the fact that it sells for less than the equivalent of ten dollars a barrel at the wellhead. If the U.S. simply returned to its natural gas consumption levels of the 1970s (before the federal government discouraged and even outlawed using the fuel) we would replace 2.5 million barrels of oil each day—about 25 percent of the oil we import.

Our domestic natural gas resources are vast. If we were to double our current consumption—using natural gas to meet 50 percent of American energy needs and thereby supplanting oil as our number-one fuel—American supplies alone would last at least 20 years. Moreover, buying additional natural gas from the enormous Mexican and Canadian reserves would allow us to secure ample energy supplies well into the next century. At the same time, we'd be cleaning up the atmosphere, since natural gas produces only two-thirds the carbon dioxide of gasoline and contains 1 percent the sulfur oxide and 10 percent the carbon monoxide of oil. In short, natural gas is better for national security, better for the environment, and cheaper for the consumer: all in all, not a bad deal for the country.

Natural gas isn't just for cooking anymore. Or just for heating or making electricity. In fact, it's poised to enter a new market: transportation. The technology for converting cars, trucks, and buses from gasoline to natural gas has been around for at least 50 years, as shown by the three-quarters of a million natural gas powered vehicles already on the world's roads. The stumbling block for conversion today is more convenience than cost: The natural gas industry offers few filling stations to compete with the ubiquitous gasoline ones. But there may be a way around this problem. After all, half of American homes are already connected to natural gas pipelines. Commuters, by using a small compressor, can tap into those lines at home and fill their tanks overnight. A flick of the switch on the open road, far from any natural gas pumps, and the car goes back to plain old gasoline.

"Natural gas is better for national security, better for the environment, and cheaper for the consumer."

Conversion is right on the verge of making economic sense. Today, it costs between $1,500 and $2,500 to turn a gasoline-burning car into a dual-fueled one. (The compressor is an additional $2,500, which your utility may cover up-front.) What are the benefits? Well, compressed natural gas is a high-powered fuel, the equivalent of about 130 octane; because it's a gas, not a liquid, it contains no particulates, which means it extends engine life by about three times the current average; and it costs *only about 70 cents per gallon.* Conversion should pay for itself within five to eight years. Even before the current crisis, United Parcel Service announced it would convert its 2,000 fleet trucks in Los Angeles to compressed natural gas. Utilities and city governments around the country—in places as diverse as Brooklyn, New York and Cannon Beach, Oregon—have already converted fleet trucks and cars. Here in Oklahoma, we run over 50 school buses on natural gas.

Because my business for the past 30 years has been the exploration and production of natural gas—and because you're an American raised on the notion that energy has to be nasty, crudish, and in short supply—you're probably a little suspicious. "If natural gas is so great," you might be thinking, "why am I still driving a gasoline-burning car? Why isn't natural gas America's chief energy source?" The answer lies in the nation's hodgepodge of energy laws, suspicion of natural gas supplies left over from the "energy crisis" of the late seventies, failure to recognize the true costs of oil, and the natural gas industry itself.

Natural Gas

For decades, natural gas production was largely a by-product of the oil industry, since natural gas reserves are found with oil. While domestic oil exploration was vigorous, the supply of natural gas was abundant, and an interstate pipeline industry developed to carry gas from the southwest to consumers nationwide. Since natural gas is delivered through pipes, local distribution companies, like electric companies, came to be regulated by the states. As a result, the natural gas industry fractured into three sectors: Natural gas was produced mainly by oil companies, transported by interstate pipelines, and sold by local distribution companies. The economic interests of the three sectors often clashed. In contrast, the oil industry is vertically integrated, with single companies controlling the entire process from production through marketing, and hence has a powerful political and economic voice.

The lack of national leadership in natural gas has held the industry back (although, for decades, it has met more than 20 percent of U.S. energy needs). For instance, the interstate pipelines and distribution companies for years battled the producers to prevent the removal of natural gas price controls that made it unprofitable to produce natural gas that was not associated with oil. Let me give you a real-life example of the damage done by those controls. Twenty years ago in Oklahoma, my company completed the then highest pressured natural gas well in the world, which was (not uncoincidentally) the second-deepest well ever drilled. The well's production of over 19 million cubic feet of natural gas per day meant it could match one-third the energy output of a nuclear plant, meeting all the natural gas needs, including industrial ones, of a town of 80,000.

But price controls administered by what later became the Federal Energy Regulatory Commission dictated that we sell our gas for the equivalent of 60 cents per barrel of oil, rendering the well hopelessly uneconomic. Can there be any wonder that natural gas producers simply stopped drilling? At the same time, the American oil fields were beginning to dry up after more than half a century of production. As the oil industry began to focus on foreign lands, the supply of natural gas associated with oil turned into a trickle.

"The doom and gloom of the Carter years led Americans to believe that energy inevitably had to be scarce and pricey."

This combination of events—the stifling of exploration and the decline of gas associated with oil fields—occurred at exactly the wrong time: during the crisis atmosphere induced by the OPEC [Organization of Petroleum Exporting Countries] shocks of the seventies. The doom and gloom of the [President Jimmy] Carter years led Americans to believe that energy inevitably had to be scarce and pricey. Despite evidence from people like me that vast resources remained, natural gas was forced to fit this pessimistic vision. Economic models forecasting the demise of the industry proliferated. The Department of Energy declared it "probable that reserves will be exhausted by the late 1980s." In 1977, James Schlesinger, then secretary of energy, responded to the notion that deregulation would leave America awash in natural gas with this comment: "I believe that's based on smoking pot." The Carter administration strove to keep despair alive by canning those at DOE and the U.S. Geological Survey who dared to be optimistic.

The Effects of Energy Legislation

In 1978, Carter signed into law the most comprehensive energy legislation ever. It contained a little good and a lot of bad. The good was that it began to decontrol natural gas prices, which led to increased exploration, which in turn led to discoveries of vast reserves of natural gas—including 300 "super wells," like ours in Oklahoma, which could generate as much power as the entire nuclear industry did in 1988. But while drillers made strike after strike in the early eighties, the market for natural gas shriveled, thanks to the bad side of the Carter legislation, which discouraged the use of natural gas in existing plants and expressly outlawed it in new industrial and power plants. In other words, just when high oil prices, public awareness of the dangers of imported oil, and the discovery of new gas reserves most favored conversion to natural gas, Uncle Sam stepped in to make conversion a crime.

"Oil has become too large an economic burden, too large a strategic risk, and too large an environmental hazard."

States, cities, and towns followed the federal lead, and soon natural gas was banned even from barbecues, outdoor lanterns, and fireplaces. As usual, it took Uncle Sam a few years to realize he'd arrested the wrong guy. Even after most of the federal bans were repealed in 1987, the perception of natural gas shortages lingers. Throughout the 1980s, the Carter legislation worked diligently to increase American pollution and reliance on OPEC. Demand for natural gas dropped from a record of 22.7 trillion cubic feet in 1972 to a low of 16.7 trillion cubic feet in 1986.

In the late eighties, oil prices stayed low enough that converting to natural gas, which requires a lot of capital upfront, didn't make economic sense for utilities and industrial plants that weren't already "dual-fueled" (that is, able to switch back and forth from oil to natural gas). Also, companies had witnessed how the federal government simply turned off the natural gas spigot only a few years before, which made them reluctant to commit to the fuel. (Americans have always distrusted natural gas, anyway: If I can't see it, smell it, or stain my clothes with it, how can I possibly run my car or my business on it?) A final weakness of the industry has been its distribution network. While you can slop oil into a truck or boat and deliver it anywhere, you can efficiently transport natural gas only through a pipeline. One million miles of pipeline crisscross the United States, but some regions of the country—in particular the Northeast and the West Coast—still can't get all the natural gas they need. Right now, almost anyone (usually competing fuel distributors with good lawyers) can delay building new pipelines by five years or more. First, they get a hearing to block construction. If that fails, they can call for a re-hearing. Failing again, they can take the case to court, beginning at the lowest level. And so on.

Congress needs to clear away the misguided regulations holding up pipeline construction, recognizing that the economic and environmental benefits of building a pipeline dramatically outweigh the costs; such red tape is now preventing natural gas from displacing another one million barrels of oil per day. Oil has become too large an economic burden, too large a strategic risk, and too large an environmental hazard. With an economically rational energy policy, we could take control of our economic future by making natural gas our principal energy source.

The U.S. Should Increase Its Use of Coal

Richard L. Lawson

About the Author: *Richard L. Lawson is president of the National Coal Association, a trade group that represents the coal industry.*

The dominant form of energy in the world today—both in the free world and the nations that are aspiring to the benefits of freedom and free economic activity—is imported oil. It is neither reliable nor low cost nor abundant.

Imported-oil dependence and energy is one of the items that will dominate our discussions and debates; it will either dominate them until solutions are found, or it will twist the future in ways destructive of progress.

During my tenure as commander of U.S. Forces in Europe in the 1980s, I had to devote considerable attention to the energy security of the United States and her allies in the free world.

The object of that attention was to protect the production and deliverability of the Mid East oil that dominates a world market that in turn dominates a free-world economy in which imports provide almost 70 percent of the petroleum consumed by the principal nations: those of Western Europe, Japan and the United States.

Imported-oil dependence is subsidized at considerable cost to U.S. taxpayers in military preparedness to respond to surprise events in the Mid East, a region of political and religious instability, and of unending surprise. . . .

The problem is not imported oil per se. The problem is in the inescapable geographic distribution of the resource: that so few control it and that so many must have it to survive. The problem is that trends in use and reserves are making the control of the few stronger.

Almost two-thirds of the proved world oil reserves are controlled by the nations of the Persian Gulf and the Mid East. About 22 percent is held by other nations of the free world and 11 percent by the socialist nations.

"The dominant form of energy in the world today . . . is imported oil."

Most Mid East producers are members of the Organization of Petroleum Exporting Countries (OPEC), but the cartel includes producing nations outside the Mid East. OPEC controls nearly three quarters of the world's oil reserves. The Mid East controls 85 percent of OPEC's reserves and, thereby, all actions.

Outside the Mid East and OPEC, petroleum is being used up faster than it is being replaced by discovery.

The most recent edition of the British Petroleum Company's *World Energy Review* estimated that at present rates of production the U.S. has 10 years of petroleum remaining; Western Europe, 12; Canada, 14; and the socialist nations, 15. The Soviet Union soon will have oil problems in addition to its other problems. The average life of Mid East reserves was 43 years.

Conservation and new discoveries may alter the figures, but not the trend.

The inescapable trend is toward concentration of the power to control availability and price. The inalterable fact is that the nations with control also have political and economic agendas that do not consider the best interests of the U.S. or the aspirations of the other nations of the world.

The trend can be delayed, but not diverted, as long as the world's economies and the world's economic growth depend on imported oil.

Richard L. Lawson, "Energy for America and World Economic Cooperation," speech delivered at the University of Kentucky, January 11, 1990.

America's import dependence is rising, world import dependence is rising. Increased economic activity in the United States and the world is hastening the trend and strengthening the power.

Those who use the most oil, and have the most at stake, have the least oil. Unless we change, demand across the board will be for more imported oil, not less.

America's imports began to exceed domestic production on a rather routine basis in 1989.

In the recent words of a former Saudi Arabian oil minister, "America is the best friend OPEC ever had."

Needed by all, controlled by a few, imported by most: oil has become more than a simple industrial commodity traded on a world market that is more-or-less free. . . .

In 1973, with the embargo, imported oil became a weapon of international politics, a means for exporting nations to exercise their political will. The first embargo failed but the point was made. Time is not on the side of the dependent.

"Those who use the most oil, and have the most at stake, have the least oil."

In 1979, in the crisis that followed the fall of the Shah [of Iran], imported oil was confirmed in the power to make and break economies. It set off a round of inflation and spawned economic problems that trouble the world today; this includes huge debt in the energy-poor, less-developed countries and huge trade imbalances.

Then in 1987, oil dependence proved its power to influence U.S. National Security Policy in the Persian Gulf deployment. And the U.S. got a foretaste of the high cost of deeper dependence as young Americans fought and died in the Gulf.

Imported oil has become a strategic commodity, one capable of making and breaking nations, of influencing their policies.

The inescapable fact is that in the future three of every four barrels of oil moving in world trade will be bent to the will of OPEC, which is bent to the will of the Mid East producers, who have political agendas.

Concentration of control and increased dependence are not conditions conducive to stability, to world economic growth or to mutually beneficial trade. Dependence is a weakness that invites exploitation. It is a real vulnerability. . . .

A Strategic Commodity

U.S. dependence can change the course of American economic, foreign and military policy. It can wrest vital decisions away from presidents by forcing short-term action that contradicts or even destroys long-term interests.

The situation neither fits nor favors free-market philosophies that fail to consider the strategic-political power that dependence gives imported oil.

It's past time to quit pretending imported oil is just another commodity traded on a more-or-less free market. It is a strategic commodity. Imported oil is a hand around the throat of the United States and the free world; and it is stronger than the invisible hand of Adam Smith's free-market economics.

But Adam Smith also noted that the true cost of anything is the toil and trouble necessary to obtain it.

The true cost of imported-oil dependence is far higher than the cost of breaking it, whatever the cost. It's time to quit pretending the hand isn't there. . . .

The potential price of forever having to send young Americans to the Persian Gulf to guard the oil lifelines for ourselves and our allies is too high, especially if alternatives exist or can be called into existence.

Policies that fail to recognize the true cost of imported-oil dependence reach beyond the illogical; policies that fail to search for alternatives reach towards the immoral.

In America's 268-billion tons of recoverable

coal reserves, which are the energy equivalent of l-trillion barrels of imported oil;

In America's clean coal technology for electric generation and industrial use;

In just-over-the-horizon technologies to realize transportation fuels from coal, and for other applications, the United States has the means to deal with four of the chief challenges of the future. These are concerns about the environment; the need for growth that will draw the world together in mutually beneficial economic activity and trade; America's ability to compete in that world economy; and a secure source of energy to support all of that activity.

In terms of energy, the recoverable U.S. coal reserve is larger than the world's proved reserves of oil.

By using the clean coal technologies as they become ready for commercial deployment the United States can hold down demand for oil imports.

"In terms of energy, the recoverable U.S. coal reserve is larger than the world's proved reserves of oil."

Indeed, since the crisis following the embargo of 1973, U.S. electric utilities have turned to coal to provide additional fuel in the equivalent of 3.1 million barrels of oil a day that did not need to be imported; it displaced much imported oil and also provided the means for growth. In the same time, national sulfur dioxide emissions were reduced by 25 percent; coal came to supply almost 25 percent of the Nation's energy; and the gross national product increased by 47 percent in terms of 1982 dollars.

Clearly, coal has had considerable success in moderating U.S. oil dependence while imparting economic strength and delivering environmental progress.

By advancing coal technologies that directly displace imported oil as a transportation fuel, and by using electricity to displace oil in the transportation sector, the United States can drive even more imported oil out of the domestic economy.

Improving the Environment

By making clean coal and other technology freely available to energy-poor nations, the United States can at once reduce worldwide demand for imported oil, establish energy efficiency in the economies of aspiring nations and help improve the worldwide environment.

Sulfur dioxide emissions are a major portion of the current acid rain dispute. Clean coal technology removes up to 99 percent of sulfur dioxide emissions in the generation of electric power, which means no legislation or limit on coal use is necessary to continue and speed up the dramatic and unbroken 17 year decline in sulfur dioxide emissions. All that is necessary is to provide for the unimpeded and economic deployment of the technology in the generation of electricity.

In addition, clean coal technology offers efficiency increases of up to 150 percent over conventional technology, which translates into international industrial competitiveness.

In its greater efficiency, clean coal technology can produce at least 20 percent less carbon dioxide than conventional technology, which goes to the heart of the global warming political controversy.

At this point, let me say clearly that I do not disparage the idea that the environment, as it may be influenced for better or worse by the activities of mankind, is a proper subject for government policy and action.

Indeed, everyone in the coal industry shares the public's concern about the environment. We, our loved ones and our posterity must live in this world. Nature happens to us all alike: like national economic decline, like the accompanying social disintegration, like war, the environment affects us all.

Global warming is a political controversy and a major cause in international environmental circles. Its foundation lies in the postulation that certain gases being emitted into the atmosphere are causing a warming of the earth via the greenhouse effect.

Some computer models of the earth's atmosphere have predicted a rise in the earth's temperature by 2050, but these computer models are crude and cannot make provision for all the ways in which the earth functions, or for all the things that affect its functions.

For example, there is a credible assertion that sunspots may influence climate; the computer models don't know how to handle sunspots, or oceans, or clouds, or large urban areas, or a lot of other things. To date these models have not been able to duplicate actual weather events.

"It clearly has not been established that the earth is warming, or that it will warm."

The conclusions of one leading model were tested against actual land temperatures for the period of 1950-1988, and it failed to come within 500 percent of explaining the past; it overstated what has happened by a factor of five.

In fact, refinements in computer modeling have led to a halving of the predicted temperature increases that set off the current debate.

The current debate was born in the record-setting heat and drought of 1988: the assertion was made by one man before a committee of the Senate that global warming had begun. There were immediate companion assertions and from lobbying groups that it is caused by carbon dioxide released in the combustion of fossil fuels; that it will melt the ice caps, cause coastal flooding and change the climate over wide areas of the earth; and that it can be staved off only by immediate worldwide reductions in the combustion of fossil fuels, which are the means of economic growth.

In the ebb and flow of scientific current events, there is an evolving consensus of scientists that no global warming has begun.

Indeed, in December 1989 a symposium on global warming in Washington had to be cancelled due to record-breaking cold. The nationwide cold wave caused electricity demand in several parts of the country to exceed the utilities' ability to deliver, and use reached levels not expected until the mid-1990s. Voltage reductions, power outages and rolling blackouts were not uncommon.

It clearly has not been established that the earth is warming, or that it will warm; that activities of man now seen by theory as driving elements will be, in fact, causative; or that all the signs some interpret as indicators of warming are not, in fact, natural and normal fluctuations. Indeed, just a few years ago similar variations in temperature were being read as signs of a new ice age.

It is established, however, that the problem, if there is one, is global; that carbon dioxide is the least radiative of half-a-dozen man-made gases thought capable of trapping heat; that emissions of other so-called greenhouse gases are growing more rapidly; that these gases are many times more radiative; that carbon dioxide contributes less than one-half to the so-called greenhouse effect; that worldwide emissions from coal-fired electric generation contributes no more than eight percent of the total global production of greenhouse gases; and that U.S. coal-fired electric utilities account for no more than three percent of worldwide production of greenhouse gases.

The Wrong Target

Nevertheless, there is serious domestic and international pressure to cap and reduce fossil fuel use to reduce carbon dioxide production that may, or may not, cause a warming that may, or may not, happen.

The first thoughts to be pressed forward in the debate are the regular and almost-ritualized responses that the organized pressure groups have to most economic activity; in this case it runs in a chain: carbon dioxide, economic activity, coal, caps, limits, carbon tax.

Every organized environmental pressure group driving this controversy would be happy to attack the three percent contribution of coal-fired electric utilities and let go the other 97 percent that isn't understood; that they don't know how to get at.

They don't know how to get at automobiles, and may fear the public reaction if they really try; but if carbon dioxide is the real concern, automobiles must be considered in ways other than raising gasoline mileage requirements: for automobiles are high on the list of aspirations of most people and every nation of the world. They don't know how to get at methane from belching cattle and rice paddies; but methane is one of the strongest and most rapidly increasing greenhouse gases, and these are among its major sources.

On one hand, there is much we don't know about the speculated-upon global warming in addition to whether it will happen. On the other, we do know that one result of caps and limits and punitive taxes is certain: economic activity of the kind that draws the world together will be curtailed dramatically.

Clean Coal Technology

This is where energy technology, especially clean coal technology, becomes very important.

Because it can produce more energy at lower levels of carbon dioxide emissions, it provides time to apply painstaking science to determine if there is a global warming without resort to caps, curbs and limits. And, finally, if science indicates there is a global warming problem, it provides time to establish other, complementary technologies to deal with it.

Clean coal technology speaks to the public's concerns while benefiting the globe: it will contribute to the world's environmental and economic prospects in a positive way.

Clean coal technology, if deployed in the developing nations that have only coal, can alter the fears about future leaps in carbon dioxide release that are part of the concern. Nations such as China and India will use their coal whether the world likes it or not; they have to. The three-billion-person increase in world population forecast by the year 2010 will demand its use.

"Clean coal technology . . . will contribute to the world's environmental and economic prospects in a positive way."

Reliance on this technology and American coal by the developing nations that have no energy but imported oil can give them a solid, secure and competitive basis for growth; and it can alter the projections for carbon dioxide release in the future.

The Soviet Union and the nations of Eastern Europe are struggling along with the most-polluting, least-efficient economies in the world.

If made available to these nations, clean coal technology would deliver impressive benefits to the global environment by dramatically lowering emissions from their industries at lower costs, which would also give them a basis to begin competing in the market economy of the free world.

If deployed logically in the United States, and encouraged in the industrialized nations, clean coal technology would go a long way towards establishing independence of action and flexibility in energy; and, once again, significantly reduce carbon-dioxide projections from coal-fired facilities.

Coal has kept 3.1 million barrels a day of imported oil out of the American economy, and it has influenced world demand for oil imports. With new technology this flexibility can be increased and expanded to other nations around the world who now must have oil to survive; and whose dependence is increasing the make-or-break power of the organized oil exporters.

U.S. clean coal technology is an alternative to imported oil that favors the environment and economic growth.

Technology Can Increase Oil Production

James Critchfield

About the Author: *James Critchfield is a Middle East expert and president of Gulf Futures Inc., which specializes in energy research.*

The era of oil—conventionally produced oil—may end during the lifetime of children born since 1985.

Not too far into the next century, conventionally produced oil will be available in meaningful quantities only in the Middle East. And not too long after that it will not be available anywhere.

So much for the bad news. The good news is that there's a great deal of oil around in forms that are just more difficult and more expensive to produce. Slowly we are learning how to extract this locked up oil at costs that may be tolerable a few decades ahead.

What many people don't realize is that there is a whole spectrum of petroleum resources beyond the various grades of crude oil and natural gas that sometimes make cameo appearances in the news media. On the heavy and thick end, this spectrum runs to heavy oil, bitumens, shale oil, and tar sands.

Most of the public is also unaware of the huge amount of conventional crude oil that is left behind when oil fields are abandoned as uneconomic. Current recovery rates average only about 33% or 34%. That means that some two-thirds of the oil in reservoirs we have tapped remains behind in the earth's strata.

In some fields, though, the recovery rate has been upped to 50%. And therein lies a major hope for the future. To use an improbable metaphor, it's as if mankind has been a very sloppy user of toothpaste tubes, not knowing how to roll them up, and is now shifting to a toothpaste pump, which squeezes more out of the reservoir.

As we will see in a moment, it is in these two areas—(1) the unconventional heavy oil part of the spectrum and (2) unrecovered conventional oil—that we will be able to extend the life of the Oil Era by at least several human lifetimes.

Those extra generations of time may be needed to perfect the full array of alternative energy systems for the planet—from photovoltaic (sunlight-into-electricity) to fusion (the sun-imitating process of fusing hydrogen atoms). And to perfect conservation, a more efficient business and home life-style.

First let's look at conventional oil, what's been thought of since the 1850s as the backbone of the Oil Era of human history.

Ultimately recoverable conventional oil is now estimated to total about 2,000 billion barrels. That figure is arrived at by taking the current recovery rate (33%) and applying it to the 6,000 billion barrels of this kind of relatively "easy" oil in the ground at the beginning of the Oil Era.

"The era of oil—conventionally produced oil—may end during the lifetime of children born since 1985."

One-third of this 2,000 billion barrels, or about 625 billion barrels, has been produced already—pumped out of the ground and consumed in the past 140 years. Global "proven reserves" are estimated at about 1,000 billion barrels (of which about two-thirds is in the Persian Gulf area). "The rest" of the conventionally recoverable oil (no one says it is *exactly* the remaining 375 billion barrels) is yet to be discovered. "The rest" we deduce to be there through sophisticated geologic mapping of the world's sur-

James Critchfield, "Stretching Earth's Oil," *World Monitor: The Christian Science Monitor Monthly,* September 1990. Reprinted with permission.

face and subsurface.

But these figures need not be the end of the history of the Oil Era. Advancing earth science research, and the technology developed from it, are slowly increasing the percentage of oil that can be recovered from old as well as new fields. Ultimately, more than 2,000 billion barrels of *extra* oil may be recovered by improved conventional methods. To accomplish that, we will have to go back into abandoned fields and force old oil to the surface. And we will have to double our average recovery rate for new oil to about 66%.

Oil History

But even that is not the end of oil history.

In addition to those extra 2,000 billion barrels of conventional oil wrested from the ground, there are some 8,000 billion barrels of extra heavy oils, bitumens and tar sands. These are petroleum resources collectively described as "unconventional."

When we look at these, the balance of oil power shifts from the Middle East (which will have a position of almost total dominance as we near the end of the conventional oil age) to the Western Hemisphere. Some 77% of unconventional oil forms are in the Western Hemisphere; 43% are in North America. Canada and Venezuela are the major sources.

The world will be less likely to come to a disastrous conflict over oil if it conserves energy and if the great concentrations of heavy oils and bitumens in the Western Hemisphere are developed in a time frame that reduces the world's projected near-dependence on Middle Eastern exports.

The Soviet Union also has about 100 billion barrels of potentially recoverable heavies and bitumens. Mikhail Gorbachev and his successors should have a keen interest in the West's developing technology for their extraction.

Efforts to find and produce unconventional oil from the heavy end of the spectrum are only beginning. But efforts to tap the two-thirds of conventional oil that is currently left behind are already under way.

The blanket name for the latter techniques is "enhanced oil recovery" or simply EOR.

EOR methods aim to flush more oil out of sub-surface reservoirs by using heat (usually steam), miscible fluids ("dry" fluids that mix with crude and coax it out of rock strata), and chemicals (which change the molecular properties of the crude oil). These techniques are sometimes used in combination with the injection of gas or water, standard methods for assisting Mother Earth in applying natural subterranean pressure to force oil to the surface.

Since the first oil shock in 1973-4, the oil industry has been conducting pilot EOR tests in virtually all regions of the world. Most successful so far have been thermal techniques. The Duri field in Indonesia, where Caltex is the operator, is now producing 150,000 barrels per day from the world's largest steamflood operation. The Kern River field in California, a thermal EOR operation, is producing 125,000 barrels per day. The Sultanate of Oman on the southeastern coast of the Arabian Peninsula at the entrance to the Gulf, has had almost a decade of experience with both thermal and chemical EOR techniques.

EOR production is more expensive than conventional production but appears economically feasible at the modestly higher oil prices that are certain to prevail in the next decade or two.

"Efforts to find and produce unconventional oil from the heavy end of the spectrum are only beginning."

Other engineering advances are also adding to our ability to tap heretofore unavailable oil. Among them: horizontal drilling; more exact computer imaging of seismic test results, which makes drilling more accurate; deeper offshore drilling that will soon allow, for instance, the tapping of large salt dome formations in the Gulf of

Mexico.

Today EOR techniques are mainly useful in getting more out of conventional oil fields, from which 95% of all usable oil still comes. But EOR methods are also spilling over into the recovery of unconventional oil substances.

The Oil Era

In short, we already have, or are rapidly developing, the technology to stretch the Oil Era well beyond its presumed obituary date. But the average person driving a car, flying on business, flipping a light switch, turning up the thermostat, cooking, or watching TV is remarkably unaware that the Oil Era—and relatively cheap energy—might end within the lifetime of his or her children. So publics aren't pushing their governments to support belt-tightening policies needed to extend the Oil Era, speed conservation, and encourage alternative energy development. And the politicians who know what is needed are often afraid to inform the public.

Oil industry specialists assume that there will be a doubling of real oil prices early in the next century. That should result in an adequate supply of energy in the United States at least through the first quarter of the century. But the economic costs will be very high. Maintaining national economic growth (and a standard of living comparable to today's) will be a painful challenge to the American people. It will be all the more difficult because of the growing realization that there is a heavy environmental cost to consuming hydrocarbon resources.

To cure that environmental problem as quickly as possible means speeding engineering research leading to more energy-efficient factories, offices, and homes. It means further improving wall and window insulation. It means less energy-wasteful production of steel and cement. It means faster development of non-polluting or less-polluting energy sources (running the gamut from photovoltaic roofs to improved-safety nuclear plants).

"Extending the life of the Oil Era remains humanity's best insurance policy against a declining standard of living."

But, while all those advances are being perfected, extending the life of the Oil Era remains humanity's best insurance policy against a declining standard of living. We need to hang onto the trapeze we're riding until we get a firm grip on the next trapeze.

The Greenhouse Effect Does Not Justify Cutting Fossil-Fuel Use

Robert James Bidinotto

About the Author: *Robert James Bidinotto writes on environmental and legal issues for the monthly magazine* Reader's Digest.

In the summer of 1988, one of the century's worst heat waves gripped the East Coast and had Midwest farmers wondering if the Dust Bowl had returned. On June 23, at a Senate hearing on global climate change, James Hansen, a respected atmospheric scientist and director of NASA's [National Aeronautics and Space Administration] Goddard Institute for Space Studies, gave alarming testimony. "The earth is warmer in 1988 than at any time in the history of instrumental measurements," he said. "The greenhouse effect is changing our climate now."

Hansen's remarks touched off a firestorm of publicity. A major news magazine speculated that the Great Plains would be depopulated. On NBC's "Today" show, biologist Paul Ehrlich warned that melting polar ice could raise sea levels and inundate coastal cities, swamping much of Florida, Washington, D.C., and the Los Angeles basin. And in his recent book, *Global Warming,* Stephen Schneider of the National Center for Atmospheric Research imagined New York overcome by a killer heat wave, a baseball double-header in Chicago called because of a thick black haze created by huge forest fires in Canada, and Long Island devastated by a hurricane—all spawned by the "greenhouse effect."

Robert James Bidinotto, "What Is the Truth About Global Warming?" Reprinted with permission from the February 1990 *Reader's Digest.*

In Paris in July 1989, the leaders of seven industrial democracies, including President George Bush and British Prime Minister Margaret Thatcher, called for common efforts to limit emissions of carbon dioxide and other "greenhouse gases." To accomplish this, many enviromentalists have proposed draconian regulations—and huge new taxes—that could significantly affect the way we live. Warns Environmental Protection Agency head William Reilly: "To slow down the global heating process, the scale of economic and societal intervention will be enormous."

"Climatic forces interact in poorly understood ways, and some may counteract warming."

The stakes are high: the public could be asked to decide between environmental catastrophe and enormous costs. But do we really have to make this choice? Many scientists believe the danger is real, but others are much less certain. What is the evidence? Here is what we know:

What is the greenhouse effect? When sunlight warms the earth, certain gases in the lower atmosphere, acting like the glass in a greenhouse, trap some of the heat as it radiates back into space. These greenhouse gases, primarily water vapor and including carbon dioxide, methane and man-made chlorofluorocarbons, warm our planet, making life possible.

If they were more abundant, greenhouse gases might trap too much heat. Venus, for example, has 60,000 times more carbon dioxide in its atmosphere than Earth, and its temperature averages above 800 degrees Fahrenheit. But if greenhouse gases were less plentiful or entirely absent, temperatures on Earth would average below freezing.

Because concentrations of greenhouse gases have been steadily rising, many scientists are concerned about global warming. Researchers at the Goddard Institute and at the University of

East Anglia in England foresee a doubling of greenhouse gas concentrations during the next century, which might raise average global temperatures as much as nine degrees Fahrenheit.

What is causing the buildup? Nature accounts for most of the greenhouse gases in the atmosphere. For example, carbon dioxide (CO_2), the most plentiful trace gas, is released by volcanoes, oceans, decaying plants and even by our breathing. But much of the *buildup* is man-made.

CO_2 is given off when we burn wood or such fossil fuels as coal and oil. In fact, the amount in the atmosphere has grown more than 25 percent since the Industrial Revolution began around 200 years ago—over 11 percent since 1958 alone.

Methane, the next most abundant greenhouse gas, is released when organic matter decomposes in swamps, rice paddies, livestock yards—even in the guts of termites and cud-chewing animals. The amount is growing about one percent per year, partly because of increased cattle raising and use of natural gas.

Chlorofluorocarbons (CFCs), a third culprit, escape from refrigerators, air conditioners, plastic foam, solvents and spray cans. The amount in the atmosphere is tiny compared with CO_2, but CFCs are thousands of times more potent in absorbing heat and have also been implicated in the "ozone hole."

No Relationship

What does the ozone hole have to do with the greenhouse effect? For all practical purposes, nothing. Ozone, a naturally occurring form of oxygen, is of concern for another reason. In the upper atmosphere it helps shield us from ultraviolet sunlight, which can cause skin cancer. In 1985, scientists confirmed a temporary thinning in the ozone layer over Antarctica, leading to a new concern: if ozone thinning spreads to populated areas, it could cause an increase in the disease.

The ozone hole appears only from September to November, and only over the Antarctic region, and then it repairs itself when atmospheric conditions change a few weeks later. It also fluctuates: in 1988, there was little ozone thinning.

Ozone is constantly created and destroyed by nature. Volcanoes, for example, can release immense quantities of chlorine, some of which may get into the stratosphere and destroy ozone molecules.

"Many scientists are troubled when inconclusive evidence is used for political advocacy."

But the most popular theory to explain the appearance of the ozone hole is that man-made chlorofluorocarbons release chlorine atoms in the upper atmosphere.

Despite thinning of upper atmospheric ozone over Antarctica, no increase in surface ultraviolet radiation outside of that area is expected. John E. Frederick, an atmospheric scientist who chaired a United Nations Environment Program panel on trends in atmospheric ozone, has dismissed fears of a skin-cancer epidemic as science fiction. "You would experience a much greater increase in biologically damaging ultraviolet radiation if you moved from New York City to Atlanta than you would with the ozone depletion that we estimate will occur over the next 30 years, he says.

Will destruction of forests worsen the greenhouse effect? When trees and plants grow, they remove CO_2 from the air. When they are burned or decay, they release stored CO_2 back into the atmosphere. In nations such as Brazil, thousands of square miles of tropical rain forests are being cleared and burned, leading many to be concerned about further CO_2 buildup.

Worldwide, millions of acres are planted with seedling trees each year, however; and new studies reveal that there has been no reliable data about the impact of forest destruction on global warming. Research by Daniel Botkin and Lloyd Simpson at the University of California at Santa Barbara and by Sandra Brown at the University of Illinois at Urbana shows that the carbon content of forests had been vastly overestimated, suggesting that deforestation is not as great a source of CO_2 as was once thought.

Can we be certain that global warming will occur? Virtually all scientists agree that if greenhouse gases increase and all other factors remain the same, the earth will warm up. But "the crucial issue," explains Prof. S. Fred Singer, an atmospheric scientists at the Washington Institute for Values in Public Policy, "is to what extent other factors remain the same." Climatic forces interact in poorly understood ways, and some may counteract warming.

At any given time, for example, clouds cover 60 percent of the planet, trapping heat radiating from its surface, but also reflecting sunlight back into space. So, if the oceans heat up and produce more clouds through evaporation, the increased cover might act as a natural thermostat and keep the planet from heating up. After factoring more detailed cloud simulations into its computer models, the British Meteorological Office recently showed that current global-warming projections could be cut in half.

Oceans have a major effect upon climate, but scientists have only begun to understand how. Investigators at the National Center for Atmospheric Research attributed the North American drought in the summer of 1988 primarily to temperature changes in the tropical Pacific involving a current called El Niño—not to the greenhouse effect. And when ocean currents were included in recent computerized climate simulations, the Antarctic Ocean didn't warm—diminishing the likelihood that part of its ice sheet will break up and add to coastal flooding.

How heat travels through the atmosphere and back into space is another big question mark for the global-warming theory. So is the sunspot cycle, as well as the effect of atmospheric pollution and volcanic particles that can reflect sunlight back into space. Such factors throw predictions about global warming into doubt.

So what is the bottom line? Has the earth begun to heat up? Two widely reported statistics *seem* to present a powerful case for global warming. Some temperature records show about one degree Fahrenheit of warming over the past century, a period that has also seen a noticeable increase in greenhouse gases. And the six warmest years globally since record keeping began 100 years ago have all been in the 1980s.

As for the past decade, the increased warmth in three of its hottest years—1983, 1987 and 1988—is almost certainly associated with El Niño events in the Pacific.

Temperature Records

Paradoxically, the historical records of temperature change do not jibe with the greenhouse theory. Between 1880 and 1940, temperatures appeared to rise. Yet between 1940 and 1965, a period of much heavier fossil-fuel use and deforestation, temperatures dropped, which seems inconsistent with the greenhouse effect. And a comprehensive study of past global ocean records by researchers from Britain and M.I.T. [Massachusetts Institute of Technology] revealed no significant rising temperature trends between 1856 and 1986. Concludes Richard Lindzen of M.I.T.'s department of Earth, Atmospheric and Planetary Sciences, "The data as we have it does not support a warming."

"Oceans have a major effect upon climate, but scientists have only begun to understand how."

Taking everything into account, few climatologists are willing to attribute any seeming warming to the greenhouse effect. In May 1989, 61 scientists participating in a greenhouse workshop in Amherst, Mass., declared that "such an attribution cannot now be made with any degree of confidence."

Is there any other evidence of global warming? Atmospheric researchers use complex computer programs called General Circulation Models (GCMs) to plot climate change. But a computer is no more reliable than its input, and poorly understood oceanic, atmospheric and continental processes are only crudely represented even in the best GCMs.

Computer calculations do not even accurately predict the past: they fail to match historical greenhouse-gas concentrations to expected temperatures. Because of these uncertainties, Stephen Schneider says in *Global Warming*, it is "an even bet that the GCMs have overestimated future warming by a factor of two."

In time, the computer models will undoubtedly improve. For now, the lack of evidence and reliable tools leaves proponents of global warming with little but theory.

Not Enough Evidence to Act

Should we do anything to offset the possible warming up of the globe? Fossil fuels now provide 90 percent of the world's energy. Some environmentalists have advocated huge tax increases to discourage use of coal and other fossil fuels. Some have suggested a gasoline tax. There are also proposals that the government subsidize solar, windmill and geothermal power; that some foreign debts be swapped for protecting forests; and that worldwide population growth be slowed.

The buildup of greenhouse gases is cause for scientific study, but not for panic. Yet the facts sometimes get lost in the hysteria. Stephen Schneider confesses to an ethical dilemma. He admits the many uncertainties about global warming. Nevertheless, to gain public support through media coverage, he explains that sometimes scientists "have to offer up scary scenarios, make simplified, dramatic statements, and make little mention of any doubts we might have." Each scientist, he says, must decide the "right balance" between "being effective and being honest. I hope that means being both."

"Further research and climatic monitoring are certainly warranted."

The temptation to bend fears for political ends is also ever present. "We've got to ride the global-warming issue," Sen. Timothy Wirth (D., Colo.) explained to a reporter. "Even if the theory is wrong, we will be doing the right thing in terms of economic and environmental policy."

But many scientists are troubled when inconclusive evidence is used for political advocacy. "The greenhouse warming has become a 'happening,'" says Richard Lindzen. To call for action, he adds, "has become a litmus test of morality."

We still know far too little to be stampeded into rash, expensive proposals. Before we take such steps, says Patrick J. Michaels, an associate professor of environmental sciences at the University of Virginia, "the science should be much less murky than it is now."

Further research and climatic monitoring are certainly warranted. If the "greenhouse signal" then emerges from the data, we can decide on the most prudent course of action.

Chapter 2

Is Nuclear Power a Viable Energy Alternative?

Nuclear Power: An Overview

John Greenwald

About the Author: *John Greenwald is a staff writer for* Time, *a weekly newsmagazine.*

Editor's note: Nuclear power is an important energy source for the United States. In 1991, nuclear power plants produced 21 percent of the nation's electricity. In addition to being a plentiful source of energy, nuclear power does not emit greenhouse gases, lead to oil spills, cause acid rain, produce smog, or create dependence on unreliable foreign oil supplies, all of which are problems caused by the burning of fossil fuels. For these reasons, proponents argue that nuclear power is becoming more attractive as an alternative energy source.

Yet critics believe nuclear power's dangers far outweigh its benefits. Two nuclear plant accidents in particular focused the public's attention on the dangers of nuclear energy. In 1979, the Three Mile Island nuclear power plant in Pennsylvania suffered a series of errors which partially melted the reactor's nuclear fuel rods, permanently incapacitating the unit. There were no significant releases of energy and no one was injured. Seven years later the Chernobyl nuclear power plant in the Soviet Union exploded, spewing a cloud of radiation that caused 23 deaths and led authorities to permanently evacuate the region surrounding the plant. Radiation from Chernobyl continues to cause illness and death in the Soviet Union and parts of Eastern Europe.

The public is also concerned about the radioactive wastes generated by nuclear power plants. These wastes remain dangerously radioactive for thousands of years and must be stored with great care. However, there is no permanent waste disposal program in the United States. The thousands of tons of waste generated by nuclear power are currently being held in special containers near reactors.

The viewpoints in this chapter provide contrasting opinions on what role, if any, nuclear power should play in meeting America's energy needs. The following overview by John Greenwald explores the renewed debate surrounding nuclear power as the United States attempts to shape an energy policy for the 1990s and beyond.

Nuclear power. The words conjure first the hellish explosion at Chernobyl that spewed a radioactive cloud across the Ukraine and Europe in 1986, poisoning crops, spawning bizarre mutant livestock, killing dozens of people and exposing millions more to dangerous fallout. Then the words summon up Three Mile Island and the threat of a meltdown that spread panic across Pennsylvania's rolling countryside seven years earlier. From these grew the alarming television programs, the doomsday books, the terrifying movies, even the jokes (What's served on rice and glows in the dark? Chicken Kiev.) Could any technology survive all that? It seemed this one couldn't. U.S. utilities ordered their last nuclear plant in 1978—and eventually canceled all orders placed after 1973. Nuclear power looked as good as dead.

Yet it lives. More than that, it is reasserting itself with great force. A survey of high-level policy leaders and futurists by Yankelovich Clancy Shulman, released in April 1991 shows a sudden upsurge in support for nuclear power following a decade of rejection. As the world worries about global warming and acid rain, even some environmentalists are looking a bit more kindly on the largest power source that doesn't worsen either problem: nuclear. New reactor designs would make accidents like Chernobyl and Three Mile Island impossible, or so the engineers say, and while much of the public is skeptical, some scientists are persuaded.

The sometimes theoretical debate is becoming intensely practical. As electric companies around the U.S. warn of periodic brownouts, people wonder, Where will we get more juice?

Nuclear power has a long way to go before it becomes the answer to that question. The public

John Greenwald, "Time to Choose," *Time*, April 29, 1991.

is afraid of it. Wall Street doesn't even want to hear about it. Most environmental groups are still virulently antinuclear. Yet here, there, in more places every day, support is building. The National Academy of Sciences called in April 1991 for the swift development of a new generation of nuclear plants to help fight the greenhouse effect. The new atomic plants already on the drawing board would replace power stations that burn coal and oil, fossil fuels that belch heat-trapping carbon dioxide—the primary greenhouse gas—into the atmosphere.

Many scientists applauded the findings of the independent academy, which conducted a 15-month federally funded study of the greenhouse problem. Says Ratib Karam, director of the Neely Nuclear Research Center at Georgia Tech: "Nuclear energy is now the only major source of power that does not produce CO_2. In terms of global society, nuclear power plants are essential."

"Americans remain deeply ambivalent about nuclear power."

Even before the academy released its report, George Bush put forth an energy plan in February 1991 that proposed greatly speeding up the procedure for licensing the new generation of nuclear plants. That is critical: public challenges to plant construction have stretched out licensing to as much as 20 years and raised building costs to such intolerable levels that many utilities have been forced to abandon plants before they ever opened.

To speed the process further, the Administration wants Westinghouse, General Electric [GE] and other suppliers of nuclear plants to build them to a standard design that would be relatively simple to repair and maintain. France, which generates 75% of its electricity from the atom—more than any other nation—has used a standard reactor since the mid-1970s, enabling any nuclear engineer or plant operator to work on 52 of the country's 55 plants at a moment's notice. By contrast, each of the 112 U.S. nuclear plants, which produce 21% of the nation's electricity, was custom built at its site. So when something goes wrong, a specialist has to fix it, causing delays that tend to make U.S. plant shutdowns longer than in France.

The new push for atomic power gained impetus from the gulf war, which focused attention on America's appetite for Middle East oil. Nuclear advocates have long argued that atomic plants could help wean the U.S. from risky reliance on energy from one of the world's most volatile regions. The effect would be small. Most utilities have already phased out their oil-fired plants, which generate just 6% of U.S. electricity and represent about 3% of the country's overall use of oil. But nuclear proponents insist that new atomic plants would further reduce America's dependence on foreign oil, enhancing U.S. energy security while reducing polluting emissions of CO_2.

The threat of climatological change could lead to a rapprochement between the nuclear power industry and U.S. environmentalists, long bitter foes. As they prepared to celebrate the 21st anniversary of Earth Day, leading environmentalists had the specter of global warming much on their mind. "Nuclear has a proven track record of producing large amounts of energy," says Douglas Bohi, director of energy at Resources for the Future, a Washington-based research group. "But the industry has to convince the public that the new technology will be safe and pose fewer problems."

Nearly everyone agrees that this challenge will be key. It will surely be one of the most daunting public relations assignments of the century. After nearly 40 years of living with the so-called peaceful atom—once expected to make electricity "too cheap to meter"—Americans remain deeply ambivalent about nuclear power. A 1991 *Time/CNN* poll conducted by Yankelovich Clancy Shulman found that 32% of the 1,000 adults surveyed strongly opposed building more nuclear plants in the U.S. vs. just 18% strongly in favor. So do Americans hate nukes? Not necessarily.

When asked which energy source the U.S. should rely on most to meet its increased energy needs in the next decade, a surprising 40% of respondents picked nuclear power, far surpassing the 25% who chose oil and the 22% who named coal.

The apparent contradiction results from the old not-in-my-backyard syndrome. Many people want nuclear power as long as it's generated elsewhere. Fully 60% of respondents said a new nuclear plant in their community would be unacceptable, vs. 34% who said it would be acceptable. Coal got a warmer reception. Only 41% considered a new coal plant in their community unacceptable, while 51% said it would be acceptable.

"A nuclear power accident anywhere stirs public fears about nuclear plants everywhere."

Such tangled feelings about the risks and rewards of nuclear power fit a worldwide pattern. In March 1991 the governments of Britain, France, Germany and Belgium—Europe's largest users of nuclear energy—jointly reaffirmed their commitment to the atom and pledged to cooperate in the development of new reactors. Yet while the statement recognized "the environmental benefits" of nuclear power and noted that it provides "one appropriate response to the challenges now confronting the entire planet," the signers warned that future development of atomic energy "must take place in conditions of optimum safety, ensuring the best possible protection both for populations and for the environment."

Safety is a vital global issue. A nuclear power accident anywhere stirs public fears about nuclear plants everywhere. Executives of U.S. utilities shuddered in February 1991 when the failure of a valve caused the worst mishap in the 20-year history of Japan's atomic power industry, crippling a plant in the town of Mihama, about 200 miles west of Tokyo. "When the skill and discipline of the Japanese falter," says Lawrence Lidsky, an M.I.T. nuclear engineer, "that means anyone can screw up."

The strongest motive for a U.S. nuclear renaissance is America's galloping demand for electricity. The Department of Energy [DOE] says the country will have to raise its present generating capacity of 700 gigawatts—or 700 billion watts—another 250 gigawatts by 2010. That is the equivalent of 250 large coal or nuclear power stations. The need will grow more acute as existing nuclear plants, which were designed to last 40 years, are dismantled and buried. By 2030, DOE says, the U.S. will need 1,250 more gigawatts of generating capacity than it has now.

The hottest argument in energy circles focuses on the right mix of fuels and conservation methods to satisfy this proliferating need for plug-in power. The issue is not whether the U.S. has enough coal. Even if the nation chose to meet all its staggering demand with its most popular fuel for generating electricity, coal, its reserves would last many decades. The question is whether America wants to bear the costs and effects of burning all that coal or would prefer the costs and effects of splitting some atoms instead.

Or perhaps it would rather do something else entirely. Environmentalist call for harnessing such renewable resources as wind and solar power and retrofitting homes and offices to use electricity more efficiently. The only trouble is that, according to the National Academy of Sciences report, "alternative energy technologies are unable currently or in the near future to replace fossil fuels as the major electricity source for this country. If fossil fuels had to be replaced now as the primary source of electricity, nuclear power appears to be the most technically feasible alternative."

That endorsement marks one of the few recent positive developments for an industry that has been mired in misery for more than two decades. Faced with an endless round of challenges, U.S. utilities have walked away from 120 nuclear plants since 1974—more than all the plants now in operation. In New York State, the Long Island Lighting Co. gave up on its completed $5.5 billion Shoreham nuclear facility in 1989 after local authorities refused to approve the firm's plans for an evacuation route for nearby residents in the event of a serious accident. The state now plans to buy the plant for a token $1—and to spend about $186 million to dismantle it.

Such fiascoes have for years discouraged virtually every U.S. utility from even looking sideways at nuclear power. "We have no plans to build a nuclear plant," says Pam Chapman, a spokeswoman for Indiana's PSI Energy. The troubled company is still reeling from the financial crisis that sandbagged it in 1984, when it wrote off $2.7 billion in construction costs for a half-built reactor. Concurs Gary Neale, president of nearby Northern Indiana Public Service Co., which scrubbed a barely started nuclear plant in 1981: "We're not antinuclear, but given the size of our company, I just don't think it ever would be practical for us."

"We as a nation should be hell-bent for efficiency. . . . We have a huge potential for savings with already existing technology."

Nor is nuclear power currently practical for any other firms in America, Wall Street experts argue. "The first utility that announces plans to build a new nuclear reactor will see its stock dumped," warns Leonard Hyman, who watches electric companies for Merrill Lynch. Hyman estimates that abandoned U.S. nuclear projects have generated some $10 billion of losses for the utilities' stockholders. "Investors are not quite ready to warm up to nuclear power just yet," says Hyman. "They're still recovering from their first chilling experience—and it was very chilling." He adds, "There is no demand for new plants, because no one wants to spend the next 10 years in court or being picketed."

All that resistance stems from fear, and the overriding fear these days is of nuclear waste. Says I.C. Bupp, managing director of the Massachusetts-based Cambridge Energy Research Associates and a long-time student of nuclear energy: "There will be no nuclear renaissance until a waste-disposal programs exists that passes some common-sense test of public credibility and acceptability."

The public's dread centers on the radioactive elements that remain in spent fuel rods after atomic reactions. While such highly toxic fission products as strontium 90 and cesium 137 have half-lives of only about 30 years, other intensely radioactive substances like plutonium will endure for tens and even hundreds of millenniums, and are piling up fast. High-level waste—that which is most radioactive—from U.S. power plants is not voluminous. More than 30 years' worth totals 17,000 tons, a thimbleful compared with the slag that would result from generating equivalent power by burning coal. Yet this waste threatens to fill all available storage space at generating facilities, and the U.S. has made little headway in developing a safe final resting place for more of it.

In 1988 Congress selected Yucca Mountain in a remote part of southwest Nevada as the site for a permanent underground repository. The state has fought the plan in a series of court battles that have helped delay the scheduled opening of the site to 2010. The DOE is meanwhile compiling a library of 10 million computerized documents that will attempt to analyze every aspect of the site to be sure it can safely hold the waste.

In light of all the turmoil, most people might be surprised to learn that a number of scientists say the waste problem can be solved with little fuss. The spent fuel rods can be buried in steel canisters thousands of feet below the surface, and experts can predict with a high degree of probability that a site will remain stable for hundreds or thousands of years. But as the public perceives nuclear waste, that's just not good enough. While the risks of so-called deep geologic disposal appear no greater than many others that Americans accept every day—crossing the street, driving a car—no scientist can guarantee that a disposal site will remain unchanged for tens of thousands of years or that groundwater may not seep into the containers at some point during the eons that the waste will remain radioactively hot. As long as the American public demands ironclad assurance that the waste cannot ever escape its containers, people's fears can never be entirely soothed.

In France, where the state runs the nuclear plants, the public seems less fearful of nuclear waste. The French convert their high-level waste into a stable, glassy substance and store it in concrete bunkers at plant sites while experts study where to dispose of it permanently sometime early next century. "The most important thing to remember is that we have time to make a proper decision," says Bernard Tinturier, director of strategic planning for the government's Commissariat for Nuclear Energy. French scientists are considering four locations around the country, including clay deposits about 120 miles north of Paris and a shale site near the Loire valley. If the French seem calmly deliberate about the issue of nuclear waste, that may be because they view atomic power as a necessity rather than an option. With virtually no oil and little coal or natural gas, France has decided to rely on its rich uranium deposits as the primary source of fuel for its power plants. The country is pressing ahead with plans to construct seven new nuclear plants by the end of the decade.

"The U.S. must decide just how practical and sensible nuclear power—and other sources of energy—really are."

With new nukes out of the picture in the U.S., utilities have been scrambling to find other sources of the electricity they need to prevent summer brownouts and blackouts that hit when demand for air conditioning peaks. To handle the load, utilities have quietly placed orders in recent years for enough gas-fired generators to produce 30,000 megawatts of electricity—equivalent to 30 large nuclear plants. But gas has drawbacks as a long-term alternative to nuclear energy. Though far cleaner burning than coal, it is still a fossil fuel that emits at least some CO_2. Reliance on natural gas would require augmenting pipelines that link the energy-rich U.S. Southwest to the populous North and Northeast, an expensive undertaking with its own environmental hazards.

So utilities are turning with increasing vigor to other nonnuclear energy sources. California's giant Pacific Gas & Electric gets a substantial 14% of its generating capacity from renewable energy sources such as the sun and wind. Its neighbor, Southern California Edison, joined forces with Texas Instruments in a six-year, $10 million project that will use low-grade silicon instead of more expensive higher grades to make photovoltaic cells that convert sunlight into electricity. Says Robert Dietch, a Southern Cal Edison vice president: "This has the potential to be the type of breakthrough technology we've all been looking for in the solar industry."

An alternative energy source that will not become practical for a long time, if it ever does, is nuclear fusion, which can use ordinary water as fuel. The difficulty is that fusion requires temperatures as high as hundreds of millions of degrees Celsius, and scientists have been unable to develop reactors that can handle that. Reports that some researchers achieved "cold fusion" at room temperature now produce more chuckles than heat.

The most productive nonnuclear, nonfossil power source in the long run may be not some new way of generating more electricity but new ways of using less. Instead of spending money to build plants, utilities sometimes find it more economical to offer customers financial incentives to use power more efficiently. In New York City, for example, Consolidated Edison spent more than $8 million in January and February 1991 on rebates to customers who traded in their energy-hogging air conditioners and lighting fixtures for efficient new models. Notes John Dillon, a Con Ed assistant vice president: "The cleanest megawatt is the megawatt not consumed."

Most environmentalists emphatically endorse conservation as a superior alternative to nukes. "Over the past decade, the U.S. has gotten seven times as much new energy from savings as from all the net increases of energy supply," asserts Amory Lovins, director of research at Rocky Mountain Institute in Snowmass, Colo. "Efficiency is a clear winner in the market, leaving everything else in the dust." Declares Lester Brown, president of the Washington-based Worldwatch Institute: "We as a nation should be hell-bent for efficiency. The exciting thing about conservation is, we have a huge potential for savings with already existing technology."

Other experts argue that the U.S. will profit from both conservation and nuclear power. "Conservation has tremendous potential," says Cambridge Energy's Bupp. "We have every reason to applaud the effort. But it will take time and good management to get the full results." Meanwhile, he says, the nuclear power industry has "invested $1 trillion over the past 30 years making plants simpler, cheaper and safer. Nuclear power should continue to provide about 20% of U.S. electric generation over the next century because it does work."

That moderate proposal seems sensible, but it won't be easy to realize. No matter how much scientific support the stricken industry receives, it hasn't a hope of getting back on its feet without lots of help from Washington, and for the moment that looks uncertain.

Utility executives must be persuaded that ordering nuclear plants again can make economic, environmental and practical sense. The first challenge, already addressed in the Administration's proposal, will be to streamline the licensing process, which now requires a set of public hearings before a plant can be built and another before it can start operating. In the case of New Hampshire's $6 billion Seabrook nuclear power station, the second round of hearings kept the completed plant idle for three years, costing its owner, Public Service Co. of New Hampshire, an extra $1 billion in interest and other expenses before the facility finally opened in 1990. To prevent such costly delays, the White House wants to accelerate licensing by compressing the two sets of hearings into one while still allowing for public comment before a plant starts up.

"Utility executives must be persuaded that ordering nuclear plants again can make economic, environmental and practical sense."

But that proposal seems sure to set off a furious battle in Congress that will test the depth of George Bush's commitment to nuclear power. "Congress is risk averse," says a House staff member. "The public doesn't like nuclear energy, and it doesn't want the right of a public hearing taken away." A careful reader of the public mood, Bush has so far shown little willingness to put up much of a fight for his program. Even chief of staff John Sununu, a former engineer who pushed hard for Seabrook when he was New Hampshire's governor, has shown at least as much interest in blocking opponents of nuclear power from key jobs in the Administration as in promoting nuclear energy.

While the White House has dithered, the DOE has invested more than $160 million in recent years to help develop a new generation of advanced reactors with standardized designs. Participants in the program include GE and Westinghouse, which have put up a total of $70 million. Washington wants four designs ready for utilities to choose from by 1995. "The key is getting the first one built," says William Young, an assistant DOE secretary for nuclear energy. That would "let the public know what it can expect."

But the question remains: Who would buy such a plant? Wall Street experts say the most likely customers could be consortiums rather than individual firms. "The next generation of nuclear reactors will be partly owned by manufacturers as well as by utilities," says Barry Abramson of Prudential Securities. "Utilities want to spread the risks around this time." That seems to be happening already. Without much fanfare, for example, Westinghouse and Bechtel, a San Francisco-based engineering firm, have formed a joint venture with the Michigan utility Consumers Power to purchase and operate nuclear plants.

The federally run Tennessee Valley Authority [TVA] could be another deep-pocketed customer for the first new reactor. TVA chairman Marvin Runyon says he may order a nuclear plant by the end of the decade. TVA also plans to restart one of three nuclear reactors at its Browns Ferry plant, near Athens, Ala. The facility had a serious fire in the mid-1970s and shut down in 1985 to correct safety problems. Runyon likes atomic energy because it is clean, but he lists four conditions that must be met if nukes are to regain the public's trust: "One-step licensing, standardized designs, a nuclear-waste-disposal program and a bold spirit of confidence."

That will be a tall order for a fractious industry that seems to have a knack for making things difficult for itself. Case in point: while some congressional law makers want to sponsor a demonstration project that would showcase new nuclear technologies and help streamline licensing procedures, squabbling manufacturers have been resisting the idea. Companies that have developed new technologies argue that they don't need the project to prove that their designs are efficient and safe. Firms whose plans are still on the drawing board are worried that the project would leave them out in the cold.

The bickering has left legislators shaking their heads. Bennett Johnston, a Louisiana Democrat who chairs the Senate Energy Committee, says he may drop a provision to fund demonstration projects from a bill he has co-sponsored to speed up the licensing of nuclear plants. Sighs a frustrated Senate staff member: "This is a hard industry to help."

It certainly is. Of all the genies unleashed by modern science, none has inspired more anxiety than the power of the atom. As if that were not disquieting enough, the industry has long been plagued by what Victor Gilinsky, an outspoken former member of the Nuclear Regulatory Commission, has called "too many deep-dish thinkers," who believed the future belonged to nuclear power and often overstated its potential. "It became a way of life instead of just a practical way of generating electricity," Gilinsky says. "The whole thing just became too ponderous, instead of practical and sensible."

"This is a hard industry to help."

Now the U.S. must decide just how practical and sensible nuclear power—and other sources of energy—really are. Nukes worry the public far more than they worry scientists who have studied their technology, yet the decision must be a matter of public will. Would Americans rather run the risk of a worldwide rise in temperatures or take the chance that steel canisters filled with high-level radioactive waste might someday leak? Or would they prefer to minimize both risks in favor of heavy reliance on efficiency and alternative energy—and then not be sure the lights will come on when they flick a switch?

The choice should not seem anguished. But following any course will require years of commitment—and as projections of electricity demand soar, there is no time to lose.

Is Nuclear Power a Viable Energy Alternative?

Yes: Nuclear Power Is an Important Energy Alternative

Nuclear Power Is an Excellent Energy Source
Nuclear Power Can Reduce U.S. Dependence on Foreign Oil
Nuclear Power Can Be Environmentally Safe
New Plant Designs Can Make Nuclear Power Safe

Nuclear Power Is an Excellent Energy Source

Frederick Seitz

About the Author: *Frederick Seitz is president emeritus of Rockefeller University in New York City, the chairman of Scientists and Engineers for Secure Energy, and the former head of the National Academy of Sciences.*

America's electricity demands are soaring, up five percent a year—twice what anyone forecast even in 1985. . . .

In fact, our demand for electricity is climbing so fast that over the next decade U.S. generating capacity must increase by a third. Fossil fuels supply nearly three-quarters of this energy. But the smoke-belching stacks of coal-, gas- and oil-fired plants are also responsible for about half of our air pollution.

That, we used to think, is a small price to pay for progress. But there is an alternative, one that produces no smoke and can actually create more fuel than it consumes. In many regions it is even cheaper than coal-fired electricity: nuclear power.

Already nuclear power is the second largest source of our electricity, and a new family of "fail-safe" nuclear reactors—some now under construction in Japan—may one day make nuclear power even cheaper and more plentiful. But before these can be built in the United States, major changes must be made in the way nuclear plants are financed, licensed and operated. For that to happen, all of us need to understand the truth about issues long clouded by misinformation.

Excerpted with permission from "Must We Have Nuclear War?" by Frederick Seitz, *Reader's Digest*, August 1990.

Getting electricity out of uranium is not difficult to understand. Each atom in this heavy metal contains a bundle of protons and neutrons held together by a powerful force. Occasionally, an atom spontaneously splits, releasing part of the binding energy as heat. It also kicks out a few neutrons that, if they strike other uranium atoms, will cause them to fracture and spray more neutrons, and so on.

In a nuclear power plant, the resulting chain reaction is kept within safe limits by sealing the uranium into zirconium-alloy tubes that are placed far enough apart to let most of the neutrons escape without hitting other atoms. The temperature of the core is controlled by inserting or removing the tubes. To make electricity, the heat from the core boils water and drives a steam turbine that cranks a generator—just as in a coal-fired plant.

The only major difference between nuclear and conventional plants is that nuclear fuel is far more radioactive. For this reason, the core must be sealed from the outside environment—and so must the spent fuel, which remains radioactive for years.

If other types of power didn't present equal or worse problems, it would make no sense to consider nuclear power at all. But they do:

"[Nuclear power] produces no smoke and can actually create more fuel than it consumes."

Coal is much dirtier than it used to be. U.S. reserves of clean-burning anthracite are virtually exhausted. Today, power plants must use soft coal, often contaminated with sulfur. When the smoke from this coal is dissolved by precipitation, it results in "acid rain."

Burning coal produces carbon dioxide as well, which can act as a blanket, trapping solar heat in our atmosphere. Eventually, this could contribute to global warming, the greenhouse effect, though there is no conclusive evidence that this has begun.

Coal also contains a surprising amount of radioactive material. Indeed, a coal-fired electric plant spews more radioactive pollution into the air than a nuclear plant.

Oil and natural gas are too scarce to meet our electrical needs now, let alone in the next century. We already import over 40 percent of our oil from abroad, and that will likely increase.

Solar power seems to be a wonderful idea: every square yard of sunshine contains about 1000 watts of inexhaustible energy, free for the taking. The trouble is, the taking *isn't* free. To meet our electrical needs, we'd have to build enough collector plates to cover the state of Delaware. No serious student of solar power expects it to be anything but a supplement to conventional electricity for decades.

Wind power generated a lot of excitement in the early 1980s, when magazines featured photographs of a "wind farm" at Altamont Pass, Calif., with hundreds of windmills. Everyone seemed to forget that taxpayers' money helped buy the farm. Today, the giant blades spin productively only half the year, because winds frequently aren't strong enough to cover costs.

Hydro power is the cleanest practical source of electricity. But in the United States, most rivers that can be profitably dammed already are.

Other, more exotic energy schemes would harness ocean tides and waves, nuclear fusion (the process that powers the sun) or heat from the earth's crust or the sea. But even proponents admit that none of these will become a major source of energy soon.

Some people believe we can solve our problems through conservation. But even if we instituted every form of conservation known, we would buy perhaps a decade before demand overtook supply again.

Advantages of Nuclear Power

Now let's look at the advantages of nuclear power.

1. *It's clean.* Radioactive emissions are negligible, much less than the radioactivity released into the air naturally from the earth or produced by cosmic rays. Standing next to a nuclear plant, I am exposed to only one-half of one percent more radiation than when sitting in my living room. A coal station, on the other hand, requires huge dumps of fuel and ashes that menace the environment.

"Nuclear waste is *not* a technical problem."

Despite a widespread misconception, nuclear waste is *not* a technical problem. The 108 nuclear plants in the United States generate less than 4000 tons of fuel waste each year. In fact, all 33 years' worth of the nation's spent nuclear fuel would only fill a football field to a depth of five feet. Non-nuclear hazardous waste, by contrast, totals 275 *million* tons *annually*. And nuclear waste is easy to monitor and control. The spent fuel can be kept on the premises for years until it decays to a radiation level suitable for trucking to long-term storage sites.

"What if a truck has an accident?" anti-nuclear activists ask. The answer is that when you're moving a chunk of waste the size of a bushel basket, it's easy to build an indestructible container. These already exist. In tests, they've been rammed by speeding locomotives and burned at 1475 degrees Fahrenheit in jet fuel without producing a single leak.

"What if the truck is hijacked by terrorists who want to build an atomic bomb?" Answer: this would be pointless. To make a bomb, the usable portions of nuclear-fuel waste would need to be extracted in a reprocessing plant costing hundreds of millions of dollars. A terrorist group with that kind of money could far more easily mine the ore, which occurs naturally throughout the world.

The best place to store nuclear waste is deep underground, because the earth provides excellent shielding. The waste will be packaged in metal-and-ceramic containers for easy retrieval.

As its radioactivity ebbs, future generations may want to recycle the unspent fuel. (Reprocessing of spent fuel is being done in Europe and Japan.) The federal government has already selected an underground storage site in Yucca Mountain, Nev. Actual use, however, must wait for years of federal testing. Meanwhile, the location is being challenged in court by the state of Nevada.

2. *It's inexhaustible.* U.S. uranium reserves will last many decades, and our long-term supply is guaranteed. Through a process called "breeding," a reactor can convert uranium into plutonium—an even better fuel. Breeder reactors, now in use in France, could thus extend the reserves for *millions* of years.

3. *It's secure.* Because it needs so little fuel, a nuclear plant is less vulnerable to shortages produced by strikes or by natural calamities. And since uranium is more evenly scattered about the globe than fossil fuels, nuclear power is less threatened by cartels and international crises.

4. *It's cheap.* In France, where nuclear power supplies 70 percent of the electricity, nuclear power costs 30 percent less than coal-fired power. This enables France to export electricity to its neighbors. In Canada, where nuclear power supplies 15 percent of the electricity, Ontario Hydro has proposed building ten more nuclear reactors over the next 25 years.

The Decline of Nuclear Power

In the early days of nuclear power, the United States made money on it too. But today opponents have so complicated its development that no nuclear plants have been ordered or built here in 12 years.

The decline of the U.S. nuclear-power industry began in 1979 with the Three Mile Island 2 accident. Radioactive emissions were negligible; the reactor shut itself down, and no lives were lost. Control-room designs are safer now, and the number of instructors and simulators at training sites has dramatically increased. Ironically, the Three Mile Island 1 plant has the best performance record in the world.

The greatest fear of nuclear-power opponents has always been a "meltdown": the reactor core overheats and breaks its seal, leaking radioactive gases. This is partly what happened at Chernobyl in 1986, and critics still cite Chernobyl as proof of what can happen here. But they fail to mention that the Soviet reactor had an appallingly obsolete design, one that is not used in the United States for generating electricity.

"An expanding nuclear-power program is vital to our economic well-being."

Today, the chances of a meltdown that would pose a threat to U.S. public health are very slim. But to even further reduce the possibility, engineers are testing new fail-safe reactors that rely not on human judgment to shut them down but on the laws of nature.

One type, first designed in Sweden, has a pool of boron-laced water surrounding the reactor core. If the core overheats, the hot water naturally rises, pulling up more boron from below, which stops the reaction in its tracks. As long as the law of gravity holds, a meltdown can't occur.

Another type, being developed at Argonne National Laboratory in Idaho, is called the Integral Fast Reactor (IFR). It recycles most of its wastes into more fuel. "It's the next best thing to a perpetual-motion machine," says physicist Charles E. Till. Where a conventional reactor might require 200 tons of fuel, the IFR would need only one ton. And because its liquid-sodium cooling system does not cause the corrosion and cracking that wears out conventional reactors, its lifetime is almost limitless.

The U.S. Department of Energy is also enthusiastic about the IFR. Says Jerry D. Griffith, Associate Deputy Assistant Secretary for Reactor Systems, Development and Technology: "If we can generate all U.S. electric power exclusively with IFRs, the atmosphere would essentially be free of pollution from utilities."...

Standardized plant designs are also necessary. No two facilities in the United States are alike, because blueprints are continually subject to NRC [Nuclear Regulatory Commission] revision. "In France," says Chauncey Starr, "they get a design approved and duplicate it all over the country. This saves a *tremendous* amount of money." Indeed, engineering work accounts for ten percent of a typical nuclear plant's cost. The NRC has agreed to approve a few off-the-shelf plans that can be used anywhere. . . .

James J. O'Connor, head of Commonwealth Edison of Chicago, one of the largest U.S. utilities, says, "An expanding nuclear-power program is vital to our economic well-being." When nominating Adm. James Watkins as Secretary of Energy, President George Bush made his support for this energy source clear. "I am convinced we are not going to solve the national energy needs through hydrocarbons alone," he said. "We must safely use nuclear power."

Splitting the Atom

I am one of the few people alive today who attended the famous conference in 1939 where Danish scientist Niels Bohr announced the splitting of the atom. The news stunned us all. Here was a process that could release a million times more energy from a lump of fuel than any fire. Some of us observed that it might usher in a golden age of global prosperity. It still can happen.

Nuclear Power Can Reduce U.S. Dependence on Foreign Oil

Bill Harris

About the Author: *Bill Harris is the senior vice president of the U.S. Council for Energy Awareness, a trade association representing the nuclear energy industry. The following viewpoint is from a speech delivered by Harris to the American Legion Annual Convention, August 25, 1990, shortly after the Iraqi invasion of Kuwait.*

The Persian Gulf contains most of the oil reserves to fuel the industrialized world. Whoever controls that oil . . . controls western economies . . . and destinies. The U.S. cannot let a regional bully wield that kind of power.

It's easy to hate Saddam Hussein. It's very easy to blame him for our involvement in Mideast madness. But we need to be honest. He had some help—unintentional though it was—right here in the United States. No one in the Middle East got us hooked on oil. No one over there put our heads in OPEC's [Organization of Petroleum Exporting Countries] noose by importing half of all the oil we use. Most important, no one over there kept us from developing energy that's Made in the U.S.A. We did that to ourselves.

It's not as if we didn't have warning. Twice before, unrest in the Persian Gulf has caused oil shocks in the West. In 1973, as you may well remember, with the Arab oil embargo. And in 1979, after the Iranian Revolution. Yes, we were warned . . . and warned again . . . that foreign oil addiction was bad for America's health. But we ignored the risk. And now our young men and women are in Saudi Arabia, a hair trigger away from war.

What more will it take for us to kick the foreign oil monkey off our back . . . to develop more domestic sources of energy . . . so we'll never again risk shedding American blood to defend a vulnerable lifeline of oil.

"Using more nuclear energy and other domestic sources strengthens America strategically."

I'm here today to talk about nuclear energy. There are many important reasons it must be a major part of our energy mix. But none more important than this: it's in our national interest. The more electricity we make using nuclear energy, the less we must make with imported oil. Since the 1973 Arab oil embargo, nuclear energy has saved this country over 4 billion barrels of oil . . . enough to fuel every motor vehicle in America—cars, buses, trucks, you name it—for well over a year. And nuclear energy has kept $125 billion U.S. dollars from flowing into the pockets of foreign oil producers. That amounts to almost half of America's entire 1991 budget for national defense.

Those are just the savings in the United States. Throughout the world, 426 nuclear electric plants in 26 nations displace 6 million barrels of OPEC oil each day. So you can see that nuclear energy furthers not just U.S. energy independence, but that of our allies, as well.

But there's still a ways to go toward real energy independence. Earlier I pointed out that nearly 50 percent of U.S. oil is now imported. If you think that's a problem . . . just hold on . . . because it's going to get worse. Some experts believe that by the year 2000, we could be importing more than 12 million barrels a day . . . nearly two-thirds of our total oil needs. That kind of dependence is bad enough . . . but the price tag is

Bill Harris, "Toward U.S. Energy Independence," a speech delivered before the American Legion 72d Annual Convention, Indianapolis, Indiana, August 25, 1990.

even worse: over $100 billion a year paid to foreign suppliers.

Given that frightening forecast, wouldn't you think every sector of our economy should try to cut back its use of oil? Well, the electric utility industry has cut back—dramatically. Between 1973 and 1987, utilities did a superb job of reducing their use of oil. Oil went from producing 17 percent of U.S. electricity in 1973 . . . to only about 5 percent in '87. How did they manage to do it? Most of the credit goes to the new coal and nuclear energy plants that came into service during that time, to take the place of oil. In fact coal and nuclear energy have provided 95 percent of *all* new electricity supply since 1973.

But during the 1980s, building large plants—like coal and nuclear energy—became financially risky for utilities. So they began to turn back to oil. Since 1987, utility oil use has increased 34 percent. New England now gets 35 to 40 percent of its electricity from oil-fired plants. On Long Island, it's close to 100 percent.

Does this make any sense? At a time when we're importing nearly half our oil, when we're paying $1.5 billion a week to foreign suppliers . . . when imported oil represents over 40 percent of the U.S. trade deficit . . . does it make sense to compound that problem by importing oil for electricity generation? I don't think so.

More Power Plants Are Needed

Using more nuclear energy and other domestic sources strengthens America strategically . . . there's no doubt about that. But nuclear energy also helps strengthen us economically. The U.S. economy is hungry for new electric supply. Demand for electricity is growing faster than supply. Just look around you, at all the ways we use electricity that our parents never dreamed of . . . microwave ovens . . . personal computers . . . new industrial technologies . . . like infrared and laser and robots . . . that make our industrial sector more productive than ever before.

No one knows exactly how much new electric capacity we'll need in the next 10 years. But make no mistake, there's a problem. History shows that if our economy grows by 2-3 percent a year, demand for electricity will grow at about the same pace—2-3 percent a year. That works out to between 100,000 and 200,000 megawatts of new capacity in only 10 years. That's *at least* as much electric capacity as there is today in all of California, Arizona, Colorado, New Mexico, Nevada, Idaho, Maine, New Jersey, and Connecticut.

Will those additional megawatts be there when we need them? It doesn't look very good. Right now, we have only 84,400 megawatts of new capacity planned, and 57 percent of that is not yet under construction. If that's not a supply-side problem, I don't know what is.

"We will need every available generating source—including nuclear energy—to meet our nation's growing demand."

The first signs of trouble have already appeared. In the last few years, we've seen brownouts, and public appeals to conserve, all the way from New England to Florida. Electric utilities across the country report electric demand not expected till the mid-'90s or even later. There's no getting around the fact that we need to build new power plants.The last nuclear energy plant was ordered in 1978. We're starting to pay the price for that . . . in less reliable service . . . and growing dependence on foreign oil. It seems to require a crisis to alert America to its energy problems. I hope it won't take an electricity crisis to help us all wake up to our urgent need to build new plants.

It is equally important to pursue conservation and efficiency to help solve our energy problems. Nobody supports energy efficiency more than the nation's electric utilities . . . in fact, they are leading the way. Utilities have in place about 1300 programs—and spend $1 billion each year—to help their customers either use less electricity or shift our usage to off-peak hours. These programs have saved the equivalent of 21 large plants. And certainly even greater savings can be achieved in the future.

But we must be realistic. Conservation and energy efficiency alone won't solve the problem of supply. The simple fact is, the U.S. will need to produce more electricity than we're capable of producing now. Today, nuclear energy supplies nearly 20 percent of U.S. electricity. We will need every available generating source—including nuclear energy—to meet our nation's growing demand.

A Clean Source of Energy

It is encouraging to note that Americans feel strongly that we must protect our environment as we meet our energy needs. And even with the crisis in the Gulf, I believe Americans will continue to care a great deal what energy production does to our environment. Nuclear energy is one of the cleanest sources of electricity. It emits no sulfur oxides . . . no nitrogen oxides . . . no carbon dioxides . . . or other greenhouse gases.

Take away the nation's 112 nuclear energy plants, and utility sulfur oxide emissions would be five million tons a year higher. That's one-half the reduction mandated in the new Clean Air Act.

Take away our nuclear energy plants, and utility nitrogen oxide emissions would be two million tons a year higher. . . .

In France, where over 70 percent of electricity comes from nuclear energy—far more than in the U.S.—there have been even more dramatic environmental benefits. Back in 1979, when France was still heavily dependent on oil and coal-fired power, their plants emitted large amounts of nitrogen oxides and sulfur dioxide. By 1987 . . . in less than 10 years . . . the use of nuclear energy had tripled . . . and those emissions had dropped dramatically. Total pollution from the French electric power system decreased by 80-90 percent. You don't hear many French environmentalists opposing nuclear energy.

The nuclear industry has no higher priority than the safe operation of its plants. The industry was severely jolted by the accident at Three Mile Island in 1979 . . . even though no one was injured. The result of Three Mile Island was a concerted industry drive for excellence. Every aspect of plant operations was examined . . . and reexamined. Nothing was taken for granted. The independent Institute of Nuclear Power Operations was created in 1979. It examines the operation of all plants and sets and enforces high standards of training and good practices. The U.S. Nuclear Regulatory Commission, which keeps close watch on the industry, reports dramatic improvements in all areas of performance. The industry is committed to maintaining and improving those high marks.

The only deaths that ever resulted from nuclear electric power happened at Chernobyl, in the Ukraine. The Chernobyl plant was a very unique Soviet design. It lacked essential safeguards that Western plants have . . . which is why it could never have been licensed to operate in the United States. There could not be a Chernobyl *accident* in the United States . . . because there could never be a Chernobyl-type *plant* in the United States. By the way, you should know that one of the benefits of perestroika is that the Soviet Union is now fully cooperating in a worldwide network to assure safe operation of nuclear electric plants.

"Let's not wait for another crisis . . . to put America on a sound energy footing."

What about the waste from nuclear energy plants? Often, people don't realize that we're talking about a very small amount. A typical plant produces about 30 tons of used fuel each year. Contrast this with the more than 300 million tons of chemical waste created each year in our country. And no other industry has managed its waste as scrupulously as we have. Every bit of used fuel ever produced by a nuclear electric plant is safely stored—mostly on plant sites in concrete storage pools lined with steel. It has never posed a threat to man or the environment. And it will remain there, in its solid form, carefully monitored and fully accounted for, until a permanent storage facility is in place.

"Nuclear energy holds out a host of benefits."

Clearly, nuclear energy holds out a host of benefits. It can help America avoid even greater dependence on foreign oil. Nuclear energy can help meet growing demand for electricity to fuel economic growth. And it can do these things safely, without environmental damage. It's really not surprising that a broad cross-section of Americans support using nuclear energy. Over 80 percent of Americans think nuclear energy should play an important role in the new National Energy Strategy being developed by the Department of Energy. That's according to a recent, independent public opinion poll. Even more interesting, among Americans who identify themselves as active environmentalists . . . 69 percent say nuclear energy will be important in the years ahead.

That's what the American people think. And their political leaders agree. Out of 48 key Congressional votes on nuclear energy legislation during the 1980s, 41 favored nuclear energy. On the state level, the National Governors Association and the National Council of State Legislatures have enacted a number of resolutions in support of nuclear energy. So you see, anyone who tells you nuclear energy has no support . . . doesn't know what he's talking about. Nuclear energy is good for this nation . . . and Americans are smart enough to know it. . . .

Let's learn from what is happening in the Persian Gulf. Let's not wait for another crisis . . . to put America on a sound energy footing. Energy security goes hand in hand with national security. Let's become as strong . . . secure . . . and energy independent as we possibly can. We owe that to our fellow citizens and to our young people. And with the help of nuclear energy, we can do it.

Nuclear Power Can Be Environmentally Safe

Fleming Meeks and James Drummond

About the Author: *Fleming Meeks and James Drummond are senior editors of* Forbes, *a financial magazine.*

In the fight against nuclear power during the 1970s and 1980s, the U.S. antinuclear groups won. They picketed, they lobbied, they fed the media a steady diet of exaggerated horror stories. They so tied up nuclear plants in the courts and in the regulatory agencies that delays lengthened and costs piled higher and higher. Then the antinukes blamed the nuclear plant owners for their cost overruns. They effectively killed nuclear power as an alternative energy source in the U.S.

Over 100 nuclear plants have been canceled since the mid-1970s. Public Service Co. of New Hampshire was forced into bankruptcy after antinukers delayed operation of its Seabrook plant, upping the plant's cost by $2 billion. (Seabrook, issued a construction permit in 1976, was completed in 1986 but did not receive a full power license until 1990.) And Long Island's Shoreham plant has been mothballed at a cost of $5.5 billion, nearly half of which will be paid by the area's ratepayers for many years to come. All this while, from the San Onofre nuclear plant in southern California to the Indian Point facility in New York, nuclear has proved that it can coexist peacefully with the environment.

Fleming Meeks and James Drummond, "The Greenest Form of Power." Reprinted by permission of *Forbes* magazine, June 11, 1990, © Forbes Inc., 1990.

Today nuclear accounts for just 20% of electricity in the U.S. Compare this to Sweden, where the figure is 47%, and Belgium, where it is 65%. In France, 74% of electricity is generated in nuclear plants, up from under 4% in 1970.

Not that the nuclear industry didn't make its share of mistakes. The industry oversold the public on the simplicity of nuclear technology, and embarked on projects without sufficient in-house expertise to monitor them. They refused to standardize reactor designs. When public relations problems multiplied, the industry did a horrendous job of handling them.

Meanwhile, U.S. oil imports to feed conventional power plants climbed higher and higher.

But as the greens, flush from their nuclear victory, move on to new scares—global warming, acid rain, the ozone hole, the whales and dolphins—their old antinuclear rhetoric sounds discordant.

Some of the groups are beginning to concede as much. Testifying before Congress in May 1990, Jan Beyea, National Audubon Society staff scientist, recommended government funding for a new generation of nuclear reactors. Why? "As a sort of global warming insurance policy, in case the transition to renewable energy systems fails to materialize."

"Nuclear has proved that it can coexist peacefully with the environment."

There are other signs that, without fanfare, nuclear is in the early stages of a comeback. Nuclear reactor vendors are once again making sales calls on the electric utilities. The U.S. Nuclear Regulatory Commission streamlined the licensing process for new plants. Smelling opportunity, the Europeans, their own nuclear markets near saturation, are buying up American reactor makers.

Environmentalists who object to a nuclear renaissance find themselves in a difficult position.

If they argue too loudly that economic growth is bad—that everyone must give up cars and disposable diapers and go back to riding bicycles and manning (or womanning) scrub boards—they will antagonize a public that is sympathetic to their calls for clean air and pure water. But if they support fossil-fuel-fired economic growth, they must accept the likelihood of more oil spills, more acid rain, more disruption of such sacrosanct, oil-laden places as the Santa Barbara Channel and Alaska's Arctic National Wildlife Refuge.

Unless the greens are prepared to argue for fewer jobs and a lower standard of living, and thus lose support, they will have to accept some kind of power generation. Solar power and wind power and the like are pie-in-the-sky. Clean, safe nuclear power is a reality. Listen to liberal Senator Timothy Wirth (D-Colo.), who has become increasingly pronuclear: "We have an obligation to try to look toward the future. One of those future technologies is solar, one is conservation and one is nuclear."

Power Crunch

The U.S. is facing a power crunch in the 1990s. Already the power shortages are beginning to crop up. When the big freeze hit Florida in the winter of 1989/90, utilities had to resort to rolling blackouts—cutting power to particular neighborhoods intermittently for up to two hours. During the sweltering summer of 1988, residents of New England experienced brownouts. And projections show demand for electricity continuing to grow faster than capacity—over twice as fast in New England over the next decade, and nearly twice as fast in Florida.

Even factoring in plants on the drawing board that, if built, will produce 168,000 megawatts by the year 2010, the U.S. will need another 186,000 megawatts of electrical capacity. Filling the gap will require new generating capacity equivalent to a quarter of the current generating capacity in the U.S.—or almost 200 large coal or nuclear plants. This conservative estimate comes from the U.S. Energy Information Administration, an independent forecasting unit within the Department of Energy.

Will they be coal plants? Fifty-six percent of the country's electricity now comes from plants fired by coal, of which plentiful reserves still exist. But it's hard to be for coal and against acid rain. Each 1,000-megawatt coal plant annually spews out 70,000 tons of sulfur oxides—the chief culprit in acid rain.

"Solar power and wind power and the like are pie-in-the-sky."

What of oil- and gas-fired generators? The more oil is shipped, the more Valdez-type oil spills are possible. Moreover, the U.S. is already setting up the world for another round of OPEC-led [Organization of Petroleum Exporting Countries] financial instability: Oil imports are today near their 1977 high of 46.5%. And while natural gas is inexpensive right now, few experts believe it will remain so.

Solar farms? Windmills? After 20 years of costly experiment, these together contribute 1% of total U.S. electric production.

Energy conservation programs have reduced peak U.S. power requirements by 21,000 megawatts, or 3% of total current capacity. That helps, but it is only a beginning.

Quietly, nuclear is already taking the first steps toward a comeback. One important such step is the Nuclear Regulatory Commission's [NRC] streamlined licensing procedures for new nuclear facilities. Unlike big nuclear users like Sweden and France, the U.S. has required a two-step licensing process before a nuclear plant can go on-line. This regulatory system has proved an effective weapon in the hands of the antinuclear forces. Under it, utilities must face months of public hearings to get a construction permit, then run the gauntlet again when the plant is finished many years later. The two-step licensing process dramatically increases the chances of costly delays because of the public hearings and last-minute changes in design standards. As a result, in this country it can take 12 years to complete and operate a plant that the French can have up and running in 6.

The simplified procedure, issued by the NRC in 1989, provides for approval of the complete design at the beginning of the process. Under the new rules, which antinuclear groups are fiercely fighting, the owner of the new reactor would automatically be allowed to fire up the plant on completion, as long as it meets the agreed-upon design standards.

This licensing change is particularly meaningful in light of a new generation of modular 600-megawatt nuclear plants, now on the drawing boards at Westinghouse and General Electric. In theory, once these standardized designs are completed and licensed—perhaps five years down the road—a utility or independent power producer could get site approval in advance and, in effect, buy the generic plans off the shelf, license and all.

Moreover, since the design would be generic, the cost of building nuclear plants could be substantially reduced.

Westinghouse projects the cost of its new modular plant at $1,370 per kilowatt-hour, or about $825 million (not including interest costs) for a 600-megawatt plant. That cost is in line with the cost of comparable coal-fired plants now. . . .

Why would these plants come in so much cheaper than their predecessors, which cost as much as $2,500 per kilowatt (again excluding interest costs) to build? Unlike in the past, each new reactor would not have to reinvent the nuclear wheel, and many of the new reactors' components could be assembled in factories, where inspection and quality control are easy to maintain, rather than on the plant sites themselves.

Signs of Rebirth

Jerome Goldberg is head of nuclear operations at Florida Power & Light, which currently generates 30% of its energy in four nuclear plants. From December 1989 to June 1990, he says, Westinghouse and Combustion Engineering have made several presentations on the new reactors to his engineers and managers. Goldberg says FP&L's immediate plans call for building six new gas-fired generators to ease Florida's power crunch. Just the same, he's listening carefully to what the nuclear vendors are telling him.

"Eventually, there will be the recognition that shutting out the nuclear option was not a good idea," he says.

"In this country it can take 12 years to complete and operate a plant that the French can have up and running in 6."

Here's a telling indication of how interest in nuclear is building again: In its 19-year history, Greenpeace, the radical activist environmental group, never bothered to take a position on nuclear power. Greenpeace leaders say they thought the battles had already been fought and won.

But now that reports of nuclear's death appear to be exaggerated, Greenpeace is mounting its own antinuclear campaign. Says Eric Fersht, who heads the campaign: "We see a lot of signs that things are at work that need to be responded to in a big way." Judging by past Greenpeace campaigns, its antinuclear campaign will be both emotional and confrontational. Nevertheless, sparing dolphins is one thing; turning down the air conditioner on a sweltering day is something else again.

Sensing that the U.S. is becoming a potentially lucrative market for nuclear plants, in January 1990, Asea Brown Boveri, a $25 billion (sales) Zurich-based power and environmental engineering firm, paid $1.6 billion to acquire Combustion Engineering, the nuclear plant maker that built 16 U.S. plants. And in September of 1989, Framatome Group, the big French nuclear plant construction firm, paid $50 million for a 50% interest in Babcock & Wilcox

Co.'s nuclear services division. Framatome is working in partnership with the German giant Siemens to develop new reactors in the U.S.

Says Nuclear Regulatory Commission Chairman Kenneth Carr, "The foreigners recognize that the market today is in the U.S., and they're focusing on this tremendously."

Among domestic reactor makers, GE and Westinghouse have designs in the works aimed at making reactors that are even safer than today's reactors, in theory as well as in practice. The new GE and Westinghouse reactors are known as advanced passive light-water reactors. They are designed around an almost fail-safe technology, which uses gravity and natural convection to cool the reactor's core if it begins to overheat. The idea is to leave as little room for human error as possible; it was human error that led to the partial meltdown at Three Mile Island.

"Reports of nuclear's death appear to be exaggerated."

Meanwhile, the industry has focused on training programs to keep the skills of operators at current plants well honed. At Three Mile Island, for instance, operators spend one week in six in training. What's more, there are currently 73 elaborate control room simulators in use around the country (up from just a handful before the accident at 1979 Three Mile Island). These simulators, built specifically for individual plants, can run operators through dozens of emergency situations, from a loss of coolant to loss of power within the plant itself.

With the new passive light-water reactors, if the core begins to overheat, valves automatically open, flooding the core with massive amounts of water stored in tanks above the reactor. This automatic process is designed to protect the reactor from damage for three days before human intervention is necessary.

On another front, a new generation of gas-cooled reactors—preferred by the Audubon Society's Beyea and most other environmentalists who are willing to consider nuclear—is under development. Lawrence Lidsky, professor of nuclear engineering at MIT [Massachusetts Institute of Technology], notes that in theory, these gas-cooled reactors are so fail-safe that a plant could withstand the simultaneous failure of the reactor's control rods, as well as a complete failure of the cooling systems. Gas-cooled reactors are probably ten years away from commercial use, but they are on the way. . . .

Obstacles to nuclear power remain—and they are formidable. While the Nuclear Regulatory Commission has simplified the federal regulatory process, state utility commissions must also sign off on new nuclear projects. Key to getting the new plants built will be the cooperation of state utility commissions in setting up cost parameters, in advance, and guaranteeing utilities that they will be able to recover costs, so long as they stay within those parameters.

Nuclear waste, too, remains a problem, though not an insurmountable one.

From Emotion to Reason

Severe as these problems are, the need for power is immense if we are to maintain our standard of living and improve it. It may take the balance of this century, but emotion on nuclear power will give way to economic and social necessities. Realities change; perceptions change more slowly.

In 1989, Three Mile Island, the would-be icon for everything that is wrong with the nuclear power industry, was ranked by respected international trade magazine *Nucleonics Week* as the most efficient nuclear power plant in the world.

New Plant Designs Can Make Nuclear Power Safe

Paul E. Gray

About the Author: *Paul E. Gray is the chairman of the governing board of the Massachusetts Institute of Technology (MIT) in Boston.*

The American nuclear industry is its own worst enemy. By trying to push ahead with vast, costly projects that have been stalled by political opposition, it exacerbates the irrational public fears that have blocked the development of nuclear power in the U.S. Instead, utilities should be exploring a new type of nuclear reactor that recent technological innovation has put within reach: a reactor type that is environmentally sound and economically competitive.

This reactor type uses new fuels, new design methods to dissipate heat, and smaller units that can be built and tested off-site. It has excited scientists and engineers world-wide, but industry and government leaders in this country—pessimistic about the public's willingness to accept nuclear power under any circumstances—are reluctant to adopt it here. That reluctance is wrong. It is time for all of us to take a hard look at modular reactors.

It has become a commonplace to say that the nuclear industry in the U.S., is dead, and that its death looks like a suicide. The problems of Seabrook and Shoreham nuclear plants are persuasive demonstrations of that commonplace.

But oil spills, undisposable garbage, polluted beaches, and—above all—steadily increasing atmospheric pollution from fossil fuel are persuading many political leaders to review their prejudices about nuclear energy. Americans who want a clean, safe and domestically produced energy source should follow—especially because all the practical alternatives to nuclear power present grave hazards to public safety and health. The perceived risks of nuclear power are grossly overestimated and usually stated without reference to the hazards of other energy sources.

There are, however, two major problems with the present generation of water-cooled reactors. The light-water reactors, or LWRs as they are known to engineers, used in nearly all the plants in operation or under construction in the United States, place heavy demands on their builders and operators. The risk they pose to public safety is an accident involving loss of coolant that could lead to the melting of fuel elements and the subsequent release of radioactivity. The safety systems for these light-water reactors are extremely complicated. These safety systems require explicit anticipation of all possible forms of failure and they must necessarily rely on probability analysis. In a world in which probability is not widely understood, such analysis is not reassuring to most of the public. While these methods lead to margins of safety that are quite acceptable, Americans remain, for the most part, skeptics.

"The perceived risks of nuclear power are grossly overestimated."

The second problem is that light-water reactors, which are custom-made at the site, cannot be tested in advance to ascertain what would happen in a true disaster.

It is possible, however, to design and build a series of small reactors that could produce the power of a large plant. These reactors could survive the failure of components without fuel damage and without releasing radioactivity because

Paul E. Gray, "Nuclear Reactors Everyone Will Love," *The Wall Street Journal*, August 17, 1989. Reprinted with permission of The Wall Street Journal, their fuels can withstand the maximum temperatures possible under the worst of circumstances. Their design limits the power density of the reactor core as well as the actual size of the core, and exploits natural processes to remove heat and avert fuel damage in the event of a loss of coolant.

Passively Safe Reactors

Such "passively safe" reactors can be designed to suffer the simultaneous failure of all control and cooling systems without danger to the public. And their safety can be demonstrated by an actual test: a West German modular reactor has passed such tests three times.

One of the most advanced of these modular reactors is under study at the Massachusetts Institute of Technology. It is based on the West German reactor that has demonstrated its safety, but adds several technologies in which the U.S. still has a competitive industrial edge. The hot gas that leaves the reactor is used directly to spin a turbine (based on aerospace designs), which, in turn, drives a small, very high speed generator (based on power electronics). This combination results in a power generating system that is substantially smaller and more efficient than current LWR systems, which are based on steam turbines and low-speed generators.

By virtue of its inherent or passive safety features, this small, gas-cooled reactor eliminates the complex, active safety systems needed by current LWRs. The gas turbine eliminates the complex, hard-to-maintain, steam generators common both to nuclear plants and ordinary fossil-fired power plants. The result is a power plant that produces electricity not only at lower cost than nuclear reactors (an easy target), but that is competitive with the projected cost of next-generation "clean" coal-fired plants. Power from such coal generators, the Department of Energy calculated in 1986, would cost an average of 5.5 cents per kilowatt hour. Power from modular reactors can be brought to market for 4.5 cents per kilowatt hour.

These savings can be realized because the new plants will be made to a single, prelicensed design in central factories. Construction costs are estimated to be less than $1,000 per kilowatt of electricity [kwe]. Costs per kwe for the Seabrook reactor in New Hampshire and the Shoreham project in Long Island were more like $5,000 to $6,000, primarily because of long delays and extensive redesign during construction. Operating costs of traditional nuclear plants are also much higher than those of modular plants would be, because the older type require very large staffs—700 people per plant—to oversee their involuted safety systems. Modular reactors could offer much more safety with staffs only half as big.

"It is possible . . . to design and build a series of small reactors that could produce the power of a large plant."

These new plants will not only be much cheaper to build, but the added bonus of high efficiency means there will be less heat to throw away. The plants will be easier to site because they cause less damage to the local environment. And, best of all, they will not do harm to the atmosphere.

These new reactors do not eliminate the waste disposal problem, but their ceramic encapsulated fuel does simplify it. A fuel that can survive unscathed in a reactor core during an accident is obviously securely packaged for disposal under more benign conditions (albeit at the cost of a significant increase in waste volume). Many of the problems associated with the high temperature achieved by the fuel of the current generation reactors are eliminated and the potential for burial in deep geological sites is enhanced. This same feature also makes it much more difficult for the discharged fuel to be processed to produce unauthorized nuclear weapons.

Smaller, modular reactors will produce less

energy than present reactors do: 100 to 150 megawatts of electrical power output compared with 1,000 to 1,500 megawatts, but this difficulty can be overcome, if necessary, by linking together a number of small, individual power-producing modules. Since each module would be identical and centrally built, licensing could be standardized and based on full-scale testing of an actual plant. This is an enormous advantage. It would allow actual demonstration of the reactors' response to severe and demanding hazards.

With an operating risk that is virtually nil and the production of significantly less radioactivity in the environment than coal-fired electric power plants, second-generation nuclear power could be a major source of environmentally sound energy if we would only take advantage of it. The failure of the government and the nuclear industry to provide leadership in developing a second generation of power plants based on these developments has already cost us dearly.

Is Nuclear Power a Viable Energy Alternative?

No: Nuclear Power Should Not Be Pursued

Nuclear Power Is Unsafe

Union of Concerned Scientists

About the Author: *The Union of Concerned Scientists is a nonprofit organization of nearly 100,000 scientists and others concerned about the impact of advanced technology on society and the environment.*

Nuclear power is the most controversial of all the energy sources used in the United States. While some people view it as a "clean-fuel" panacea, others consider it a major threat to public health and safety. Some of the key issues in this controversy are discussed below.

There are 110 nuclear power plants licensed to operate in the US, with a combined capacity of 105,000 megawatts. These plants provide about 19% of the nation's electricity. No new nuclear plants have been ordered since 1978, and 118 plants, including all those ordered since 1974, have been canceled or deferred.

Although the cost and performance of nuclear power vary considerably from plant to plant, recent experience suggests that nuclear power has become the least competitive of conventional electricity sources. Costs of $2-$3 billion per plant are now commonplace, with some plants costing upwards of $5 billion. The average output of nuclear plants is only about 60% of designed capacity, because many plants are forced to shut down frequently for repairs and maintenance. (Coal plants run at about 80% of capacity.)

In the 1980s, the time required for construction of a nuclear reactor typically ranged from 8 to 14 years. Although regulatory delays and intervention by citizen groups are often blamed for this poor record, the real roots of the problem lie in faulty and incomplete design work, inadequate quality control during construction, an inability to secure necessary financial backing, poor management, and the nuclear industry's lack of credibility in the eyes of the media and the public. The financial woes of the nuclear industry have continued despite billions of dollars in subsidies from the federal government.

There are a number of major safety issues associated with nuclear power. Some of these include:

Premature Aging. Although nuclear plants are designed and licensed to operate for a period of 40 years, many plants may have to be closed well before their licenses expire. Exposure to levels of humidity, temperature, and radiation more extreme than expected can result in the thinning, cracking, and rupturing of pipes and the malfunctioning of vital instrumentation. Aging increases the probability that safety systems will fail when subjected to the harsh conditions of an accident.

"A host of so-called generic safety issues common to many plants . . . have never been resolved."

Inadequate Containment Design. In the event of a major accident at a nuclear plant, releases of radioactivity into the environment are supposed to be prevented by a building called a containment structure. One type, called pressure suppression, uses a pool of water or baskets of ice to condense radioactive steam released in an accident that otherwise might rupture the containment structure. Unfortunately, recent assessments by the Nuclear Regulatory Commission (NRC) indicate that this type of containment may not be up to the task. Shortly after the Chernobyl accident in 1986, a top-ranking NRC official concluded that there was a *90%* chance that the thin steel pressure-suppression contain-

From the Union of Concerned Scientists pamphlet "Nuclear Power: Past and Future." Reprinted with permission of the Union of Concerned Scientists.

ment used on 24 General Electric plants would fail during a major accident, jeopardizing the safety and lives of thousands of citizens. Other kinds of containment are not fail-safe either. A 1987 NRC assessment of three so-called dry containments (thick, reinforced-concrete buildings) concluded: "In general these data indicate that early containment failure cannot be ruled out with high confidence for any of the plants" examined.

"It is doubtful that the nuclear industry will be able to regain the confidence of the public."

Problems at Babcock & Wilcox Plants. Eight nuclear plants are based on the same design as the reactor at Three Mile Island (TMI), the site of the nation's worst nuclear accident. Uniquely sensitive to sudden changes in temperature and pressure, these Babcock & Wilcox-built plants have proved to be extraordinarily difficult to control, as evidenced by the meltdown at TMI and near-misses at the Rancho Seco plant in Sacramento and the Davis-Besse plant in Toledo. This has raised serious questions about their safety during an emergency shutdown. The NRC's own records have clearly shown that these plants are more dangerous than other pressurized-water reactors, yet many of the agency's recommendations for safety upgrades remain unfulfilled. (In response to its poor performance and safety record, the Rancho Seco plant was closed in 1989 by a citizens referendum.)

Generic Safety Issues. Despite the fact that 110 plants have been licensed to operate, a host of so-called generic safety issues common to many plants throughout the country have never been resolved. Acknowledged by the NRC but tabled for further study, the existence of these issues raises serious questions both about the safety of plants now on-line and the NRC's commitment to protect the public. Among the most serious unknowns are: the capability of safety control systems to survive fires, earthquakes, or hydrogen explosions; the capability of reactor systems to respond to an emergency-shutdown command; the integrity of the steel cladding of reactor vessels; and the extent to which human operators will respond correctly to sudden events or abnormal conditions.

Poor Siting. In order to provide "defense in depth," operators of nuclear plants are required to prepare plans for the emergency evacuation of all people living or working within a 10-mile radius. Unfortunately, several plants (e.g., Indian Point and Shoreham in NY and Seabrook in NH) have been sited within or close to densely populated areas where evacuation of large numbers of people is not feasible. Others have been sited in areas prone to earthquakes. Furthermore, when it has become clear that states and communities have been unable to develop an adequate emergency plan for a poorly sited plant, the NRC sometimes has relaxed its rules in order to allow a plant to be licensed.

Overall, the NRC estimates that there is a 45% chance that a meltdown will occur at a US reactor within the next 20 years. Although not every meltdown would lead to a containment failure putting thousands of lives at risk, some would. Beyond the potential loss of lives, the US General Accounting Office has estimated that the financial consequences could range from $67 million at a small plant in a rural area to $15.5 billion at a large plant in an urban environment.

Nuclear Waste

A typical nuclear power plant generates more than 30 metric tons of high-level radioactive waste annually, much of it in the form of spent nuclear fuel consisting of uranium and plutonium isotopes. Some of the waste will remain hazardous to humans for thousands of years and must be disposed of without leakage to the environment. Exposure to even minute amounts can cause death, cancer, or genetic disorders. About 14,000 metric tons of uranium waste are currently stored on-site in spent-fuel pools at reactors throughout the country. By the year 2000, more than half of these pools will be filled to capacity unless they are expanded, the spent fuel rods are reconfigured, or the material is shipped to other, larger sites.

"The likelihood that an existing reactor will suffer a catastrophic accident remains unacceptably great."

The Department of Energy (DOE) is investigating a site at Yucca Mountain, NV, for a permanent underground nuclear-waste repository. The examination of this site is mired in controversy and has already cost over $500 million. Because of the geologic conditions there, some scientists have questioned whether the proposed deep-underground facility will be safe. DOE recently announced that the facility cannot be opened until 2010—at the earliest. By that time almost all of the spent-fuel pools at US reactors will be full.

Future Reactor Designs

Nuclear power is often offered as a solution to the greenhouse problem, since nuclear reactors do not emit any carbon dioxide, the principal greenhouse gas, into the atmosphere. Several new reactor designs are under investigation, some of which utilize "passive" safety features to minimize the risk of a major accident. In theory, passive safety features would rely on natural forces or physical principles to keep a reactor from going out of control, rather than on mechanical or electronic measures requiring intervention by human operators.

Although some of these reactors may well prove safer than the current generation, a number of important questions remain unanswered:

• Can critical design elements (such as the fuel pellets in the new MHTGR [modular high-temperature gas-cooled reactor]) be manufactured consistently to the exacting specifications that will ensure safety features work as planned?

•Can and will the designs be "proof tested" to demonstrate that safety features are reliable—before any commercial reactors are granted a construction license?

•Will the quality of construction be any better than in the past?

•Given the track record of the last 30 years, how will the cost of advanced reactors compare with conventional power plants and renewable energy sources?

•Where and how will the additional nuclear wastes generated by new plants be disposed of?

•Which states and communities will be willing to accept the siting of additional waste-disposal facilities or the shipment of wastes across their borders?

•How can the diversion of nuclear waste suitable for use in nuclear weapons be prevented?

•Will the nuclear industry be more willing to accept stringent regulation and enforcement than it has been in the past?

Unless and until these questions are satisfactorily resolved, it is doubtful that the nuclear industry will be able to regain the confidence of the public. In the meantime, the likelihood that an existing reactor will suffer a catastrophic accident remains unacceptably great.

Nuclear Power Supporters Exaggerate Its Benefits

Nathaniel Mead and Ray Lee

About the Authors: *Nathaniel Mead is a free-lance writer who specializes in reporting on health, politics, and the environment. Ray Lee, an environmental activist, has been involved in nuclear issues since 1977.*

"Just as America gave birth to nuclear technology in the 1940s, we can lead the world into a new era of safe, reliable, economical, and environmentally clean nuclear power in the 1990s," President George Bush told the Nuclear Power Assembly. "This clean domestic source of power lessens the risk of energy dependence on foreign sources."

This glowing pro-nuclear promise—note the word "clean" in each of the two sentences—plays on some acute public fears. The Persian Gulf crisis has once again highlighted the vulnerability of foreign oil supplies. A decade of grim warnings from scientists points to excessive fossil-fuel burning as a cause of global warming. The options seem painfully clear: Either we revamp our petroleum-based economy or we face ever-spiraling fuel prices and a bloody war in the Middle East.

Against that backdrop, the energy conglomerates have mounted an intense media campaign to persuade the public and policymakers of the need for nuclear power. With prompting from a former nuclear engineer, Presidential aide John Sununu, the Bush Administration is putting its weight behind nuclear power. And despite a popular belief that the nuclear industry entered a terminal tailspin after the 1979 Three Mile Island accident, nuclear power is poised for a major revival.

The device for making this happen is "nukespeak," the use of manipulative messages aimed at achieving public acceptance of nuclear power. Nukespeak involves a calculated distortion and suppression of facts about nuclear power, and corporate control over scientific research and public information.

Spearheading the propaganda effort is the U.S. Council for Energy Awareness (USCEA). The industry's seasoned media-relations arm, known until 1987 as the Atomic Industrial Forum, USCEA sees public acceptance as the key to a nuclear comeback.

"The primary obstacle is not the industry, which has an excellent track record, but public perception," says Scott Peters, USCEA's manager of media services. The Council has called on politicians, utility managers, business executives, and university scientists to "reexamine nuclear's environmental benefits"—and spread "the good news."

"Nuclear power is poised for a major revival."

The nuclear-power industry is seizing on the Persian Gulf crisis and heightened concern about the environment. "Nothing creates a climate for nuclear better than a perception of need," says Carl Goldstein, vice president of USCEA. "If the public feels that it is needed, then they will overcome their qualms about it. With the [Gulf] crisis and concerns over global warming, the public is waking up to the benefits of nuclear. It is a clean energy form in most respects."

USCEA works constantly to determine which images and phrases incline public opinion toward nuclear power. According to the Council's official media brochure, *Mission, Methods, and*

Nathaniel Mead and Ray Lee, "Nukespeak," *The Progressive*, December 1990. Reprinted by permission from *The Progressive*, 409 East Main St., Madison, Wisconsin 53703.

Benefits, the primary function of USCEA is to "marshal research, advertising, media, and public relations . . . to draw attention to the issue, generate positive editorial comment, broaden public support, and . . . provide a more favorable business climate at the Federal, state, and local levels now and in the future."

How do Americans feel about nuclear power? Polling data gathered by USCEA in 1990 suggest that people want nuclear plants to be "safer," "new and improved," and to have "advanced" designs. People dislike such terms as "passively safe," "inherently safe," "user friendly," and "standardized designs."

Nukespeak

The key stratagem of nukespeak is to turn reality on its head, to make irrational policy seem rational. Thus, energy conglomerates now promote nuclear power as among the cleanest energy sources available. The West's economic engine will be fueled by electric power, we are told, and a new generation of nuclear plants is our only hope of doing this without dirtying the skies. Solar and wind power, nuclear's primary eco-competitors, are branded by nuclear supporters as pie-in-the-sky solutions.

As the publicists succeed in refining nukespeak, mainstream coverage of nuclear power tilts toward industry. Editors who buy USCEA's sales pitch routinely indulge in selective quoting, sensationalism, trivialization, or other kinds of misrepresentation of nuclear issues.

One of the more egregious examples is an article called, "Must We Have Nuclear Power?" by Frederick Seitz, a prominent physicist, in the August 1990 *Reader's Digest.* Seitz lavishes praise on nuclear power—the only way to achieve "cleaner air *and* economic growth." And he glosses over the problems of nuclear energy. For instance, he says of the nuclear-waste problem: "The spent fuel remains radioactive for years." Years? Try *hundreds of thousands* of years.

At USCEA, such terms as "atomic" and "power" are strictly avoided because they remind people of the Bomb and "old-generation" atomic reactors. Instead of energy efficiency, the propagandists speak only of conservation, a term most Americans associate with privation and discomfort. From every angle, the industry paints nuclear power as the bright hope of a high-tech, growth-oriented future.

Putting Up a Front

The drive to sell nuclear power, however, is not merely an exercise in linguistic subterfuge. Equally potent is the industry's recruitment of academic scientists whose credibility, in the public eye, far exceeds that of an explicitly corporate team of "experts." Whenever lawsuits or bad press pose a threat, the nuclear industry rolls out an esteemed Ph.D., invariably cited as an "independent" source, to offer opinions consistent with the industry's point of view. Not surprisingly, independent scientists whose positions oppose the industry's are either selectively quoted or never heard from. And when new studies about health risks come out, the industry twists the results to its own advantage.

In September 1990, for example, the National Cancer Institute released a major study of cancer deaths in 107 counties containing sixty-two nuclear facilities. The report acknowledged high rates of cancer in some counties and low rates in others, and concluded that nuclear power plants posed no significant hazard overall. But the report had serious flaws.

"The key stratagem of nukespeak is . . . to make irrational policy seem rational."

"The NCI study didn't look at other environmental sources of cancer, such as chemical plants, in control counties," says John Gofman, a physician and professor of medical physics at the University of California, Berkeley, and former associate director of the Lawrence Livermore Laboratory. "Since the study did not measure any releases from the facilities, there's no way to know which areas of excess cancer were due to excess radiation. It's exactly the wrong kind of study to do."

Also, "radioactive fallout could have been carried from the nuclear plants to nearby control counties, making them look relatively worse," Gofman notes.

"None of my colleagues take this kind of study very seriously," Gofman adds. "But I believe it will be used very, very opportunistically as proof that nuclear facilities don't hurt anyone."

"Independent scientists whose positions oppose the industry's are either selectively quoted or never heard from."

In its own glossy version of the report, sent out to major news media, the USCEA stated that the conclusions were not surprising to the scientific community, "which believes health risks posed by nuclear energy plants are virtually nil."

Fudging the Figures

The U.S. Government has systematically downplayed the hazards of nuclear power. As early as 1974, the *New York Times* exposed the Atomic Energy Commission's ten-year effort to suppress research findings on safety and health risks associated with nuclear energy. Since then, many critics have accused Federal officials of abusing the science of radiation and health.

In 1990, Jay Gould, formerly a Westinghouse employee and member of the Environmental Protection Agency's [EPA] science advisory board, provided evidence that the Government was running a disinformation campaign in which EPA press releases were presented as scientific reports. EPA officials have not only misrepresented the risks, claims Gould, but have frequently falsified radiation measurements taken around nuclear power plants.

Fudging figures and reconstructing data are tantamount to criminal offenses among scientists. But a more intractable problem involves optimistic interpretations of data used to determine risk—a practice less likely to be detected and rebuked. Lack of clear information and consensus on radiation-related risks leads to confusion about human safety. Because risk estimates involving radiation exposure entail a wide range of statistical uncertainty, nuclear proponents tend to focus on the lowest part of the range, which is associated with the lowest risk.

Such optimism is clearly unrealistic and dangerous. "Altogether, the ranges of estimated hazards to public health from . . . nuclear power plants are so wide as to extend from negligible to substantial in comparison with other risks to the population," writes John Holdren in the September 1990 *Scientific American*. "The very size of the uncertainty is itself a significant liability."

Gofman concurs: "The fact that there's a wide range of uncertainty in current risk-estimates does not justify a retreat to wishful thinking." The 1990 report of the National Academy of Sciences committee on the biological effects of ionizing radiation used *animal* research to justify lowering risk estimates by two- to ten-fold for low-level radiation exposure, Gofman notes. But findings from at least nine well-designed studies on *human* populations indicate a five- to thirty-fold increase in the risk from low-level exposures, which take their toll in a cumulative fashion.

Estimating Health Hazards

The most extreme instance of unrealistic optimism among the "experts" is the view that *beneficial* health effects may result from low doses of ionizing radiation—a theory known as "hormesis." In a 1989 *Science* article, Leonard Sagan argued that low-dose radiation stimulates the DNA-repair and immune systems, thereby increasing resistance to disease. Gofman devoted an entire chapter of his 1990 book, *Radiation-Induced Cancer from Low-Dose Exposure,* to a careful review of each paper cited by Sagan and other hormesis proponents. Not one of the studies offered unequivocal support for the theory.

Despite the growing popularization of such theories in the mass media, the public's concern about radiation hazards continues to bedevil the nuclear industry. While acknowledging that high-level radiation exposure does pose a threat, the industry insists that low-level radiation is safe because there's a "threshold" of exposure below which no hazard exists. This is based, in part, on the idea that the body "repairs" such damage to its DNA. But such speculative concepts as "safe threshold" and "repair" are, in Gofman's view, "a misuse of science in the service of nuclear politics."

Since the 1970s, the industry has used the term "permissible" for those releases of radiation authorized by the Government and determined by cost/benefit analysis to pose an "acceptable risk" to public health. "Safe threshold" is the underlying rationale for the Nuclear Regulatory Commission's [NRC] move to declare some low-level radwastes "below regulatory concern"—a decision that would allow nuclear facilities to dump radioactive garbage at any municipal landfill or incinerator. Since the NRC has no plans to monitor radiation levels at these disposal sites, the chances for abuse are substantial. Deregulating portions of this waste as "below regulatory concern" will save the industry an estimated $1 billion in waste management costs over the next two decades.

The ultimate costs, however, would be borne by the public. The NRC's assumption of "acceptable risk" is one incidence of cancer for every 100,000 people. Thomas Cochran of the National Resources Defense Council told *The Bulletin of the Atomic Scientists* that, by NRC logic, "it is 'below regulatory concern' to randomly fire a bullet into a crowded Manhattan street on the basis that the individual risk to a person in New York City is less than one in several million."

A New Breed

Having developed a new generation of reactors, the industry hopes to persuade utilities to order new plants—something they haven't done in more than fifteen years. The industry's main strategy is to declare the reactors "inherently safe," based on a generic, modular design that would include "natural" control features.

Though the new breed of reactors seems headed for regulatory acceptance, some seasoned engineers remain skeptical.

"The public's concern about radiation hazards continues to bedevil the nuclear industry."

"Frankly, I don't believe any of the new generation should be called inherently safe," says Greg Minor, an engineer who worked with General Electric for sixteen years before joining a nuclear consulting firm. "They may be safer in some respects, but it's misleading to say they're safe, period. You can't put enough design detail down in advance to know how well a plant is going to operate. There will be a period of learning by trial and error. Some of the new plants will be built without containment and prototypes; and the first one that goes awry in ways that aren't anticipated is going to erode public confidence immediately."

Depicting reactors as kinder and gentler is a first step toward making nuclear power seem cleaner and greener. Ever since the 1990 Earth Day, when the USCEA produced flashy posters and brochures propounding a benign nuclear ecology, the green theme has become increasingly pivotal to the industry's propaganda campaign.

Consider the *Forbes* article of June 11, 1990, called "The Greenest Form of Power," by Fleming Meeks and James Drummond. On the cover, nuclear power is referred to as "the environmentalists' best friend." The article begins by saying that since the 1970s and 1980s, when "antinuclear groups . . . fed the media a steady diet of exaggerated horror stories . . . nuclear has proved that it can coexist peacefully with the environment." The authors have apparently concluded that all is well at Three Mile Island and Chernobyl—and always has been.

Judging by appearances, of course, nuclear power *is* cleaner than fossil fuels. No one has ever seen or smelled radioactivity. Unlike coal- and oil-fired plants, nuclear generators don't produce smoke or other emissions rich in carbon dioxide. A September 10, 1990, news release from USCEA stated, "Nuclear energy is the only major source of electricity—other than hydroelectric dams—that does not emit greenhouse gases or other pollutants." No emissions—that's what two top USCEA officials told us.

But nuclear power plants are by no means a "smokeless" antidote to air pollution. In the real world, such plants routinely produce several hundred varieties of fission products, some of which are gaseous—the kryptons and the xenons. These radioactive gases are released daily from nuclear plants at "permissible" (NRC-sanctioned) levels. When USCEA says "no emissions," it really means "no greenhouse gases," or no emissions worth mentioning.

"Depicting reactors as kinder and gentler is a first step toward making nuclear power seem cleaner and greener."

The nuclear industry has rightly pegged coal as a major source of sooty smoke, smog, greenhouse gases, and acid rain, all of which are absent from nuclear energy production. But then comes the more devious claim: "A coal-fired electric plant spews more radioactive pollution into the air than a nuclear plant," as *Reader's Digest* put it in August 1990.

It's true that coal contains uranium and that coal-fired plants create radioactive emissions. But when one considers the total nuclear fuel cycle, much greater quantities of radon and other radioactive gases are emitted in the mining and milling of uranium, before the fuel even reaches a nuclear reactor. "Figuring in this initial step, the amount of airborne radioactive pollution produced in the overall operation of any one nuclear plant far exceeds that which a coal-fired plant produces," says Gofman.

Nuclear Waste

Accidental radioactive releases further undermine the industry's bold green message. The Nuclear Regulatory Commission documented more than 30,000 "mishaps" at nuclear plants in the United States between 1979 and 1987. By the NRC's own estimates, the chance of "a severe core meltdown" occurring at one of the 112 licensed U.S. nuclear plants in the next fifteen years runs as high as 45 percent. The global scenario is even more disturbing: West German and Swedish scientists predict a 70 percent chance of a Chernobyl-scale accident occurring at one of the world's nuclear plants over the next five or six years.

Many U.S. officials say that nothing like Chernobyl exists in the United States. But the Chernobyl explosion produced radioactive clouds that sent fallout as far as Japan and the United States, and Gofman calculates that almost half a million cancer deaths and just as many nonfatal cancers may eventually result from Chernobyl's fallout in Europe and the Soviet Union alone.

And the difficulty of getting rid of radioactive waste renders all talk of "clean" nuclear power irrelevant. No safe method of containing high-level waste has been found, and any attempt to curb global warming with more nuclear power would mean a corresponding increase in such radwaste.

"If 400 nuclear power plants were built (possibly enough to make some impact on global warming)," notes Scott Denman of the Washington-based Safe Energy Communications Council (SECC), "approximately five radwaste dumps the size of the proposed Yucca Mountain (Nevada) site would have to be built—an extremely difficult technical problem and a near political impossibility."

Some nuclear advocates contend that climate

changes could ultimately claim far more human casualties and cause far greater environmental damage than any nuclear accident. But posing these stark alternatives assumes that no benign solution exists.

"It only makes sense to choose an energy source that doesn't have a harmful end-product," says Gofman. "We're talking about an absurd trade-off. Compared to fossil fuels, nuclear power produces far more *lethal* pollution and is already moving toward sacrificing a safe food chain for millennia."

When nuclear power supporters apply green rhetoric to battling the greenhouse effect, they ignore the fact that the automobile is the principal source of greenhouse gases. According to SECC, substituting nuclear plants for all existing power facilities would reduce greenhouse gases by only about 5 percent. Some estimates suggest that curbing worldwide emissions of carbon dioxide even marginally would require building one new power plant every two days for the next thirty-eight years. And at that rate, the energy required for the complete process—mining and processing uranium, constructing facilities, enriching fuel, and disposing of radwastes—would conceivably result in an increase rather than a decrease in climatic devastation. . . .

"Nuclear power plants are by no means a 'smokeless' antidote to air pollution."

Economist Amory Lovins of the Rocky Mountain Institute agrees but emphasizes cost-competitiveness in the energy marketplace.

"You have to look at all the ways to get electricity and pick the cheapest," Lovins says. "You can define 'cheapest' in narrow economic terms or add in uncounted environmental costs. Either way, the cheapest way is to use the electricity we already have far more efficiently.". . .

Nuclear power is, in fact, the most capital-intensive of all energy sources. Several studies show that energy efficiency is cheaper by a factor of seven to ten than operating a typical nuclear plant, *even if the plant is built free.* West Germany and Japan, with two of the world's strongest economies, are running at twice the energy efficiency of the United States; they use half the energy to produce the same unit of gross national product.

Though centralized hard-tech energy options have dominated the media, a 1989 poll indicated that almost 80 per cent of American voters ranked funding of renewable sources—solar and wind power—as top budget priorities for the Department of Energy, and preferred by a margin of three-to-one using electricity more efficiently rather than building new power plants.

Time to Reverse Course

Recent administrations, however, have taken the opposite route. Ronald Reagan slashed R&D [research and development] for energy efficiency by 70 percent, and for renewable energy sources by 85 percent. The Bush Administration has done little to redress the imbalance and continues to subsidize nuclear power above all other energy sources.

But as oil supplies dwindle and become more costly, the economy and the political system face difficult decisions regarding sustainable energy use. Fortunately, nuclear power is not the only available alternative. When it comes to cleaner and greener energy technologies, efficiency and renewable sources represent the sane energy path—if we can compel our leaders to follow it.

Nuclear Power Is Not the Solution to Environmental Problems

Paul Hansen

About the Author: *Paul Hansen is Midwest regional director of the Izaak Walton League of America, a conservation group based in Minneapolis, Minnesota.*

Advocates for the financially troubled nuclear industry have recently been campaigning hard for renewed spending on their industry because of the alleged environmental benefits of nuclear power. Their concern is not to save the environment, but to save their dying industry.

Nuclear power cannot substantially alleviate global warming, since electrical generation accounts for only about 15 percent of our nation's greenhouse gas emissions. Even doubling nuclear capacity to generate electricity would reduce those emissions by less than 3 percent.

If we were to take the same capital that nuclear advocates would have us give their industry, and use it to provide homeowners and businesses with loans and incentives to invest in the new generation of highly efficient motors, lights, heaters, air conditioners, refrigerators, and other technologies, we would safely make about seven times more power available for immediate use. Putting these investments in nuclear power would provide no power for five to 10 years, and then only at the risk of a nuclear accident.

As much as 75 percent of the electricity produced in the United States and Canada each year is wasted through the use of inefficient motors, lights, and appliances, according to scientists at the Lawrence Berkley Lab. Similarly, Minnesota's Department of Public Service estimates that 52 percent of our electricity could be cost-effectively saved through increased efficiency.

Paul Hansen, "Efficiency Trumps Nuclear Power," *The Christian Science Monitor*, May 23, 1990. Reprinted with permission.

"Nuclear power cannot substantially alleviate global warming."

Much of the blame for the nation's massive waste of electricity rests squarely on some of the same electric utilities and public regulatory systems that are bankrolling the campaign for new nuclear plants. According to the National Association of Regulatory Utility Commissioners' Conservation Committee: "Utilities' rate structures encourage the production and sale of as much electricity as possible because selling more electricity always adds to a utility's profits, regardless of the cost of producing the extra power or the price at which it is sold; and investments in conservation measures always hurt a utility's bottom line, regardless of their cost-effectiveness."

Advances in lighting systems provide the best example of the potential for reducing pollution and energy costs by increasing the efficiency of electricity end use. A new compact fluorescent bulb, for instance, uses one-fifth the electricity of conventional lighting, provides equal or better lighting, lasts up to 13 times longer than incandescent bulbs, and provides substantial cost savings. The use of just one bulb prevents the emission of close to a ton of carbon dioxide (greenhouse gas), 21 pounds of sulfur (acid rain), a significant amount of nitrogen oxides (smog), a broad array of toxic emissions, and the solid waste associated with the 13 bulbs it replaces. New types of fluorescent tubes, occupancy sensors that turn off lights in empty rooms, and other new technologies can do even more. In contrast to nuclear power, investments in all of these technologies save the consumer many times their purchase price in electricity and labor.

On a larger scale, hot water heaters, air conditioners, other appliances, and industrial motors that do the same job as current models for a fraction of the energy are available. With the addition of weatherization, load-management strategies, and other methods, there is an enormous potential for energy and environmental savings in homes and businesses. Unfortunately, the higher initial cost and simple lack of understanding of the full benefits of efficiency have prevented widespread application of such means, even though lifetime-use costs are much lower and investments in efficiency yield an average of seven times more energy than the same investment in a nuclear power plant.

"Most of our pressing environmental and economic concerns are directly tied to energy use."

Some utilities in New England, New York, New Jersey, Maryland, Delaware, and Wisconsin are moving ahead with least-cost strategies that provide utilities with a return on investments in efficiency incentives. California leads the way, with a plan to have utilities invest $300 million in 1991, with $200 million being invested in 1990. They expect savings of $1.1 billion from this investment, not including the savings from avoided environmental impacts. The second largest northeast utility has invested $65 million, and expects a savings of $160 million, not counting environmental savings.

Efficiency First

The nuclear industry is correct on one point—most of our pressing environmental and economic concerns are directly tied to energy use. Even for those of us who do not unequivocally oppose nuclear power, investing in a nuclear future before we have taken advantage of opportunities to increase efficiency is environmentally and economically foolish. It will mean that money and energy will continue to be wasted, North America will continue to wane in competitiveness with Europe and Japan, and the environment will continue to be degraded.

The practice of giving our public trust and dollars not to the most effective technology, but to the technology with the best lobbyists, must stop if we are to address the environmental crisis.

New Plant Designs Do Not Increase Nuclear Power's Safety

Ken Bossong and John Sullivan

About the Authors: *Ken Bossong is director of the Critical Mass Energy Project for Public Citizen, a nonprofit consumer advocacy group in Washington, D.C. John Sullivan is the editor of* Public Citizen *magazine.*

The nuclear power industry is exploiting the growing concern over a number of environmental problems—from global warming to acid rain to oil spills—to promote a new generation of nuclear power plants.

Legislation has been introduced in Congress to expand federal funding for research and development of new reactor designs—an area that had been cut during the Reagan years due to the restrictions of the Gramm-Rudman-Hollings balanced budget law. Among the prominent congressional backers of expanded nuclear power is Sen. J. Bennett Johnston (D-La.), chairman of the Senate Energy and Natural Resources Committee. . . . "Nuclear power is poised to return to its rightful place as the technology of choice for electric power generation of the 21st century," Johnston told an industry trade group in May 1989. "[It] is inevitable in any workable strategy to combat the greenhouse effect." The Bush administration is also supporting an expansion of nuclear research. Testifying before a House subcommittee, a top official from the Department of Energy (DOE) observed that "development and implementation of advanced reactor technologies is crucial for the country."

Yet there is little substance to the industry's claim that engineers could design a new generation of reactors to correct the economic, safety and environmental problems that have plagued existing plants. Furthermore, most advanced reactors exist on paper only; constructing demonstration models and bringing commercial units on line could take 20 years or more—a time frame that critics say is unrealistic if nuclear power is to make a significant contribution to solving the global warming problem. As Scott Denman, director of the Washington, D.C.-based Safe Energy Communication Council, points out: "Nuclear power offers a fool's gold solution to global warming. Energy efficiency improvements and renewable energy sources are our best defense against the greenhouse effect."

The industry's most optimistic projections suggest that to build and operate new "advanced" reactors would cost at least as much as existing reactors. But before it can even launch any major initiative such as that, officials must resolve a fundamental question: Should they pursue evolutionary improvements in the design of existing light-water reactors—which use water to cool the uranium-oxide fuel rods—or shift to radically different designs, of which there are at least a half dozen.

"Nuclear power offers a fool's gold solution to global warming."

Proponents of an evolutionary strategy believe it is wiser to draw on the experience of designs that have been in use for three decades, rather than explore unproven concepts.

But the industry remains divided on that question. An official with the U.S. Council on Energy Awareness, the industry's public relations arm, has acknowledged that "the U.S. in-

Ken Bossong and John Sullivan, "A New Generation for Nuclear Power?" *Public Citizen,* September/October 1989. Reprinted with permission.

dustry knows light-water reactors" and that other designs "are untested and unneeded." But some of the leading builders of nuclear plants, such as Westinghouse and General Electric, appear to be hedging their bets and are pursuing research in both areas.

Further, the cost to design, demonstrate and commercialize new plants—whether improved, light-water reactors or those with more radical innovations—will demand a significant up-front investment. Yet many electric utilities have been so scarred by their experiences with nuclear power that they are breaking away from the nuclear option rather than encouraging it.

Thus, a nuclear revival would be virtually impossible without major government subsidies; given annual budget deficits of $100 billion or more, it is unlikely those subsidies would be adequate. Although current proposals in Congress call for expanding R & D [research and development] programs, they fall far short of the tens of billions of dollars needed.

Reactor Costs

Moreover, the smaller, "modular" reactors envisioned by some industry advocates may actually prove more expensive than the larger, present-generation plants due to economies of scale: A small reactor costs proportionately as much to run as a large one because of the same support services required, regardless of size. In fact, a recent Public Citizen study found that the two U.S. nuclear plants that are the most expensive to operate—the Big Rock plant in Michigan and Yankee Rowe in Massachusetts—are also the smallest.

One of the most widely discussed designs for the future is a modular, high-temperature, gas-cooled reactor, which uses helium as its coolant rather than water. One drawback, however, is that it produces less power for its volume than conventional reactors; in general, the more power a plant produces, the cheaper the electricity is for the consumer. . . .

Many observers also question the economics of a second reactor design that is receiving considerable attention within the industry—the Process Inherent Ultimate Safe design, or PIUS. They doubt whether the reactor can function over a long period of time, or whether it could be restarted quickly after a temporary shutdown for normal maintenance. Consequently, it may be necessary to construct additional, expensive, back-up capacity.

The Safety Problems

These are already evident in each of the new reactor designs under consideration. For example, the gas-cooled reactor is frequently referred to as "meltdown-proof," and both American and foreign companies are reportedly moving quickly to bring the design to market. Yet this so-called "immunity" to meltdown is a feature only of smaller versions of the reactor, not larger ones. Even still, the design of those smaller reactors poses serious problems. One of the reasons it is purportedly immune to meltdowns is that the design allows outside air to circulate in the reactor, which removes heat generated by the decay of radioactive materials during fission and which could otherwise melt the core. However, to increase air circulation, designers would omit the massive containment systems that characterize current commercial nuclear plants. Although there would be some type of confinement building, it would not contain any material escaping from the core during an accident. Exactly how much radioactivity could escape from the core even during normal operations is unknown.

"The smaller, 'modular' reactors . . . may actually prove more expensive than the larger, present-generation plants."

The Nuclear Regulatory Commission's (NRC) advisory committee on reactor safeguards, which has generally been sympathetic toward nuclear power, characterized this feature as "a major safety tradeoff." The committee said it was not convinced sufficient protection could be provided without a containment building designed to trap the bulk of the radiation released during an accident. The cost of such a massive structure could make a new generation of reactors uneconomical.

Potential problems also exist with other designs. A liquid-metal reactor proposed by General Atomics could experience sodium fires, leading to what the NRC advisory panel termed—in characteristic understatement—"severe consequences." Moreover, the PIUS design has a unique and potentially unstable steam generating system, and scientists have not yet studied all the possible consequences that could stem from various types of accidents. Much the same is true for a fourth design called an integral fast breeder reactor.

"Nuclear plants would actually produce more carbon dioxide . . . than renewable technologies."

Another significant drawback to a new generation of reactors is that none of the designs will solve the problems associated with radioactive waste, which many critics call "the Achilles' heel" of the nuclear power industry. Existing commercial nuclear reactors now produce some 2,000 metric tons of highly radioactive waste annually, and there is still no permanent waste site in operation. There is increasing evidence that the Yucca Mountain site, designated as a permanent repository by Congress in 1987, is unsuitable because of the potential for flooding from the water table beneath the site as well as its susceptibility to earthquakes.

Other Alternatives

Numerous studies show that a mix of investments in energy efficiency, renewable energy technologies and natural gas offers a better solution to global warming than does a resurrection of nuclear power. A recent DOE report points out that the construction, operation and fueling of nuclear plants would actually produce more carbon dioxide—the largest of the atmospheric gases contributing to global warming—than renewable technologies.

That contradicts the common perception that nuclear is a clean technology, which forms the basis of the industry's promotion of a new generation of plants. Moreover, industry officials fail to point out that the fossil fuel plants they aim to replace are responsible for only 4 percent of the global emissions of greenhouse gases.

Finally, a Public Citizen study released in May 1989, *Power Surge*, documented the growth potential for solar and other renewable sources. Those technologies already contribute more to the country's energy supply than nuclear power—8.6 percent vs. 6.9 percent—and can double their contribution by the year 2000 with the proper incentives.

Chapter 3: Preface

Should the U.S. Expand Its National Energy Strategy?

For many years, the U.S. government had no official energy policy. However, when the 1973 Arab oil embargo plunged the U.S. into its worst economic period since the Great Depression, the government began to see the need for such a policy. Americans had become too dependent on foreign oil, and the nation needed to explore new energy sources. The embargo prompted then-president Richard Nixon to formulate Project Independence, a program designed to wean the nation off oil imports altogether by 1980. This was the first formal energy policy ever formulated by an American president. The plan proposed easing environmental restrictions on the development of new energy sources, halting the switch from coal-generated electricity to oil-generated electricity, and creating the Energy Research and Development Administration. But Congress, fearing potentially high costs, rejected Nixon's plan. Instead, Congress encouraged the development of nuclear power plants and more efficient uses of coal. Although Congress's plan was much less revolutionary than the president's, it marked a new era of concern about fulfilling America's energy needs.

In 1975, Nixon's successor, Gerald Ford, continued to attempt to create a national energy strategy by proposing the Energy Policy and Conservation Act. The act called for oil price decontrol, gasoline rationing, and fuel efficiency guidelines. Ford's plan was partially approved, and Congress incorporated energy measures of its own, including requiring minimum gas mileage standards for new automobiles.

Federal energy policies were expanded again after Jimmy Carter became president in 1977. Carter proposed the National Energy Plan, which advocated energy conservation, the development of renewable energies, and the increased use of coal and natural gas. As part of his plan, Carter proposed tax credits for homeowners and businesses that invested in renewable energy. In 1978, Congress passed the National Energy Act, which included many of Carter's proposals. The act established tax credits and other incentives for the use and development of renewable energy sources such as solar heating systems. It also levied taxes on large cars that used much gasoline and offered incentives for carpooling.

With the election of Ronald Reagan in 1980, the focus on energy changed from conservation and renewable energy to finding new deposits of fossil fuels. During his eight-year tenure, Reagan proposed the decontrol of domestic oil prices, a reduction in conservation and renewable energy programs, tax incentives to encourage fossil-fuel exploration and production, and the expansion of nuclear energy use. Reagan reduced the role of the federal government in formulating energy policy by reorganizing the Department of Energy and reducing its budget. The Reagan plan achieved many of its goals. Primarily, production of nuclear power and coal increased during the Reagan years, and oil imports were reduced.

President George Bush's National Energy Strategy (NES), announced in February 1991, also advocated an expanded role for fossil fuels. Developed through more than a year of hearings and testimony conducted by the Department of Energy, the NES contains more than one hundred proposals for the future course of energy production in the United States. A main directive of the NES is to increase the exploration and production of U.S. oil deposits. To accomplish this, the NES recommends that 1.5 million acres of Alaska's Arctic National Wildlife Refuge be made available for oil exploration. The NES also urges a reconsideration of nuclear energy plants, few of which were built during the 1980s because of excessive costs and public concern about safety. These high costs were caused by increased federal regulation, including strict environmental and safety standards. To allow utility companies to build more plants, the NES recommends easing some federal restrictions, including allowing plants to operate for sixty rather than forty years and reducing the number of public hearings required before a plant is approved for operation.

The NES proposals have been met with public opposition, particularly the proposal to increase domestic oil exploration. Many environmentalists, mindful of the *Exxon-Valdez* oil spill in 1989, protest that further oil exploration in Alaska could damage the wildlife refuge. Other critics charge that since the NES seems to favor a continued reliance on oil as a source of energy, it dooms the U.S. to continued imports of foreign oil. This reliance could mean another energy crisis if foreign exporters placed an embargo on U.S. imports. As former Arizona governor Bruce Babbitt charges, "The plan perpetuates America's addiction to fossil fuels." Consumer advocate Ralph Nader similarly blames Bush for creating "a national security risk" by relying heavily on foreign oil.

However, many support the Bush plan. Texas senator Phil Gramm endorses the policy of increased domestic oil production. Gramm argues that spending more money on domestic supplies and less money on foreign oil imports would provide a much-needed boost for the economy. Louisiana senator J. Bennett Johnston agrees with Gramm: "Rather than having a one-word energy policy—a policy based on import—our policy should be one that is made in America." In addition, Bush's secretary of energy, James Watkins, contends that the environmental risks associated with oil exploration are necessary if oil imports are to be cut.

Whether or not the Bush administration policies will guide the United States toward a beneficial and productive use of energy remains to be seen. The authors in the following chapter debate whether the U.S. needs the National Energy Strategy and whether or not the federal government should assume a prominent role in energy policy-making.

Should the U.S. Expand Its National Energy Strategy?

Yes: The U.S. Should Expand Its National Energy Strategy

The U.S. Should Pursue Energy Conservation

Global Tomorrow Coalition

About the Author: *The Global Tomorrow Coalition is a group of private organizations and individuals concerned about population growth, global resources, and the environment.*

Energy is among the most essential of our Earth's resources. Without the heat, light, and food it provides, human civilization would not exist. Since World War II, the world's energy consumption has increased about fourfold. The use of fossil fuels has grown rapidly, and enabled many nations to achieve high standards of living. Yet most energy is used inefficiently, and widespread consumption of coal and oil produces pollution that threatens air quality, vegetation, and climate stability. To a large extent, the ability of our environment to support life depends on the kinds of energy choices we make, and especially how efficiently we use our energy supplies. . . .

There are vast differences between nations in how much total energy they use, how much they use on a per person basis, and how efficiently they use it. Many factors affect a nation's energy use, including population size and distribution, geography, climate, and level of development. . . .

A nation's energy consumption in relation to its economic output, sometimes called energy intensity, reflects the country's economic structure and level of development as well as how efficiently energy is used within that structure. . . . Denmark, Japan, and West Germany use about half as much energy as the United States per unit of economic output.

Other studies show that West Germany's per capita energy consumption is about half that for the United States, and that Japan spends only 5 percent of its gross national product on energy compared with 10 percent for the United States.

In the western industrial countries, since the first OPEC [Organization of Petroleum Exporting Countries] oil price increases in 1973, there have been significant improvements in energy efficiency in all major areas of energy use—including the industrial, transport, service, and residential sectors. Industrial nations used about 1 percent less primary energy in 1985 than in 1980, in contrast to the 26 percent increase in the five years before the 1973 oil price increases. And between 1973 and 1985, per capita energy use in OECD [Organization for Economic Cooperation and Development] nations dropped 6 percent while per capita gross domestic product rose 21 percent. For the first time, the link between economic growth and increased energy use was broken.

"An increase in efficiency is usually the cheapest 'source' of energy."

Unfortunately, in the last few years, momentum toward greater efficiency has been slowed by lower energy prices and the waning commitment of governments to conservation. In the United States, efficiency improvements in industry have leveled off since 1983. The U.S. government has reduced fuel economy standards for new cars and has slashed the budget for research and development of energy efficiency by 50 percent since 1983.

Improvements in the efficiency of energy use provide the quickest, least expensive means to alleviate many energy-related problems. An increase in efficiency is usually the cheapest "source" of energy. To recast the old metaphor, a kilowatt saved is a kilowatt supplied. By providing energy for actual or final uses ("end uses") such as lighting and transportation more efficiently, increased demands for such services can be met without more imported oil, coal-fired electric plants, nuclear power, or environmental pollution. In the United States, opportunities have already been identified for boosting energy efficiency that cost one-half to one-seventh the expense of new energy supply.

"Improved energy efficiency can easily be incorporated into the design of new buildings."

Regarding environmental benefits, one study suggests that a 2 percent annual increase in worldwide energy efficiency could slow the buildup of atmospheric carbon dioxide enough to keep the global average temperature to perhaps within 1 degree Celsius of current levels, thus avoiding the most serious effects of climate change. Another study shows that energy efficiency improvements are a much cheaper way to slow the carbon dioxide buildup than an expansion of nuclear power; improved efficiency displaces nearly seven times as much carbon per dollar invested.

Energy can be conserved in many ways, ranging from turning down thermostats to using more-efficient appliances and vehicles. Through improved efficiency, the amount of energy consumed in buildings, industry, transportation, and other sectors of society can be substantially reduced.

Buildings and Appliances

There is vast potential for improving energy efficiency in the heating, cooling, and lighting of buildings and in the operation of appliances such as stoves, water heaters, and refrigerators. Important savings can be made by reducing heat loss from poorly insulated buildings and replacing inefficient furnaces and air conditioners.

A call to the "house doctor" might be the best way to cut energy consumption in existing homes. In an experiment by a Princeton, New Jersey, research group, house doctors made detailed energy audits of homes. Using infrared scanners and house pressurization techniques, they could quickly identify defects in the shells of houses that were allowing heated or cooled air to escape. After leaks were sealed, fuel use was cut by an average 30 percent, and lower energy bills enabled home owners to gain a 20 percent annual return on their investment in energy conservation. Homeowners can save even more money by finding and fixing energy leaks themselves.

Windows are a major source of energy loss in commercial and residential buildings. Each year, losses through American windows equal the energy that flows through the Alaskan pipeline. Losses can be greatly reduced by installing a "heat mirror" film that lets in light while doubling insulation value, and by removing the air between the glass sheets of double-pane windows.

Energy Efficiency

Some owners of residential buildings and businesses lack the capital to make investments in energy efficiency. The cooperation of public utilities with consumers and with groups called conservation companies can solve this problem. These companies test buildings and make energy-saving improvements free of charge. To recover their investments, such firms charge a fixed monthly fee or a percentage of the money saved in energy costs. The conservation business can become more profitable and much more widely applicable with the cooperation of power companies. Some farsighted public utilities, recognizing that saved energy costs less than building new power plants, are willing to pay conservation companies for kilowatts saved. Such arrangements promote energy efficiency and provide economic benefits for all parties involved.

Improved energy efficiency can easily be incorporated into the design of new buildings. Super-insulated homes can be heated for one-tenth the average cost for conventional homes. Such houses are heated primarily by lights, appliances, and the body heat of the residents, requiring only occasional use of small space heaters. The need for costly central heating systems is often eliminated. Ventilating heat exchangers can simultaneously remove any indoor air pollution and preheat water for washing and cooking.

"Major improvements in the efficiency of energy use . . . would yield important benefits for the United States."

Compared with the average efficiency of equipment in use in 1985, the most efficient residential appliances and equipment available in the United States use much less energy. The best commercial models of refrigerators, central air conditioners, and electric water heaters use 50 to 60 percent less energy, and the best gas furnaces, water heaters, and ranges use between 26 and 43 percent less energy than 1985 averages. Improved designs now under study or development could increase these energy savings to between 59 and 87 percent compared with the 1985 average.

In the United States, about 20 percent of electricity is used for lighting. In commercial facilities, lighting accounts for between 25 and 40 percent of energy use; part of this is used to remove heat generated by the lighting. Energy used for lighting can be decreased by up to 70 percent while maintaining or improving lighting quality. Newer lighting sources also last longer than conventional ones. Standard incandescent lamps operate for about 750 hours, while newer alternatives can last for 10,000 hours while using up to 75 percent less energy.

A World Resources Institute study, The End Use Global Energy Project, estimates that a family living in a super-insulated home with the most efficient appliances currently available would consume only one-quarter the energy used by a typical household.

In the early 1970s, the town of Osage, Iowa, began an energy conservation program that has cut its natural gas consumption by 45 percent since 1974. Osage residents save energy by plugging leaky windows, insulating walls, ceilings, and hot-water heaters, and replacing inefficient furnaces. In 1988, the program saved an estimated $1.2 million in energy costs. Much of the saving has resulted from initiatives by the local utility company, which offered free building thermograms to locate heat losses, and now gives customers fluorescent light bulbs that are much more energy efficient than incandescent bulbs.

In 1986, the U.S. Congress passed the National Appliance Energy Conservation Act that established energy conservation standards for most major home appliances and central heating and cooling systems. The standards are being phased in over a five-year period and are expected to lower residential energy use 6 percent by the year 2000.

Industry

Industry has made significant gains in energy efficiency. In OECD countries, energy intensity—the amount of energy required per unit of industrial production—has fallen 30 percent since 1973. Japan has led the way, making energy-efficient industry a national priority. Full-time energy managers are required by law in all Japanese companies that use substantial amounts of energy. In Sweden, a new technology called "Plasmasmelt" is being developed that will reduce the energy intensity of their already efficient steel industry by 47 percent. Recycling scrap metals takes much less energy than producing new metals from ore. Fabricating a diecast machine part takes 95 percent less energy when recycled aluminum is used in place of primary metals. In the United States, the proportion of aluminum produced from recycled metals grew from 25 to 50 percent between 1970 and 1983.

One of the most promising developments to improve energy efficiency in industry and in cities and towns is cogeneration, a system that simultaneously produces heat and electricity or other forms of energy such as mechanical power.

For the greatest efficiency and lowest cost, the cogeneration facility should be near the site where the heat can be used.

In a typical operation, fuel (usually natural gas, but also wood, plant wastes, coal, or oil) is burned in a boiler to produce steam. The steam turns an electric generator and is then recaptured for heating, refrigerating, or manufacturing processes rather than discarded to the air. The system can more than double the usable energy obtained from each dollar invested. A utility plant producing only electricity is about 32 percent efficient; a cogenerator using the same amount of fuel can approach 80 percent efficiency. The cost of electrical capacity from cogeneration systems is usually less than half the cost of new coal or nuclear power plants.

Cogeneration can be utilized in hospitals, schools, office buildings, and apartment complexes. The amount of electricity provided by cogeneration is increasing rapidly. U.S. production will rise from 13,000 megawatts in 1985 to over 47,000 megawatts (the equivalent of 47 large nuclear plants) when projects already under way are completed.

So far, there is little accurate information about cogeneration activity in the Third World, but the International Cogeneration Society reports that some developing nations have shown substantial interest in small-scale cogeneration applications.

Transportation

Most means of transportation, including cars, trucks, and airplanes, run on petroleum products. In the United States, 63 percent of all oil is used in the transport sector. To reduce environmental pollution and dependence on unstable oil supplies, it is crucial that the transportation sector be as energy efficient as possible.

Considerable progress has been made in automobile fuel efficiency. By 1988, all U.S. cars averaged 19 miles to the gallon (MPG), compared with 13 MPG in 1973. New cars now average over 25 MPG in the United States and more than 30 MPG in Europe and Japan. However, compared with prototype, super-efficient models, most new cars are still "gas guzzlers." Prototypes now exist that are more spacious, more responsive, and safer than many current models and can get 60 to 100 MPG. They use lightweight materials and design improvements such as continuously variable transmissions. Experts have estimated that the cost of purchasing and operating such cars would be roughly the same as for the inefficient cars of today. The higher initial cost would be offset by fuel savings over the life of the car.

"Most nations have only begun to exploit the opportunities for improved energy efficiency and the potential of renewable energy sources."

In 1975, the United States enacted its most effective energy legislation, the Energy Policy and Conservation Act, which required new car fleets to improve their fuel efficiency to 27.5 miles per gallon (MPG) by 1985. This standard helped sustain efforts to improve fuel efficiency after oil prices collapsed in the early 1980s, and the average efficiency of U.S. passenger cars rose from 13 MPG in 1973 to 25 MPG by 1985. Yet in spite of this improvement, U.S. fuel economy is well below the 30-33 MPG level achieved in other industrial nations.

In 1986, following appeals by luxury-car manufacturers, the U.S. fuel efficiency standard was lowered from 27.5 MPG to 26 MPG, and by late 1987, the major U.S. car manufacturers had stopped much of their research on more fuel-efficient cars. In 1989, however, members of the

Bush administration have called for a federal fuel-economy standard of 27.5 MPG for 1990, 40 to 50 MPG for 2000, and as high as 75 MPG by 2025.

A study by the American Council for an Energy-Efficient Economy showed that major improvements in the efficiency of energy use in the transportation, industrial, commercial, and residential sectors would yield important benefits for the United States in terms of economic well-being, competitiveness, national security, and environmental quality.

Superconductivity

Looking to the future, superconductive materials are being developed that conduct electricity with no energy loss. Although not ready for commercial applications, superconductors could enable development of more-efficient electric motors and storage systems, and allow dramatic improvements in the overall efficiency and flexibility of energy use. . . .

As noted, the collapse of oil prices in 1986 has begun to slow the progress of energy efficiency. Consumers are finding that investments in conservation have longer, less attractive payback periods. To avoid the cycle of "cheap energy" leading to higher consumption and increased pollution followed by future oil price shocks, incentives are needed for continued investment in energy efficiency.

Governments can play a role in providing the necessary incentives. Estimates show that government incentive programs could triple the current $20 billion to $30 billion annual world investment in energy efficiency. Major increases in efficiency programs also could help avoid the enormous potential costs associated with global warming.

The gasoline tax in the United States is far lower than in most industrialized countries, averaging about 29 cents per gallon. In Japan, the tax is about $1.60 per gallon and in Italy, about $3.30. As a result, gasoline in most industrial countries costs two to four times as much as in the United States. Not coincidentally, drivers in American cities buy four times as much gasoline as residents of European cities. Many energy experts have recommended that U.S. gasoline taxes be increased as a means of reducing pollution and oil dependency.

A higher gasoline tax would have several related benefits. Most importantly, it would allow cleaner, more efficient means of transportation to compete fairly with motor vehicles. The current U.S. gasoline tax does not begin to cover all the costs associated with cars and trucks. Expenses such as road building and maintenance, lost tax revenues from paved-over land, traffic regulation, accidents and health care, and driver education may require up to $300 billion a year in subsidies by local, state, and federal governments.

U.S. citizens currently support motor vehicle transportation through income, property, and other taxes regardless of how much they drive. A substantial increase in the gasoline tax could distribute these costs more fairly, especially if there were compensation for low-income groups. Public demand would increase for alternative transportation, fuel-efficient cars, and community planning to reduce driving distances. Fuel-efficient cars could also be promoted through a "gas guzzler-gas sipper" tax plan in which a tax on relatively inefficient new cars would be used to provide rebates for high-efficiency models.

"Major increases in efficiency programs also could help avoid the enormous potential costs associated with global warming."

To combat global warming, a "carbon tax" could be levied on fuels in proportion to the amount of carbon dioxide they create per unit of energy. For example, coal would be taxed almost twice as heavily as natural gas. Clean energy sources would be made more competitive while those that threaten our climate would be discouraged.

At the state level, incentives for energy conservation and improved energy efficiency can be highly effective in slowing or even reversing the growth in energy use, while the absence of conservation incentives can result in rapid growth. Between 1977 and 1984, California and Texas followed different energy strategies. California set strict conservation standards for all economic sectors, but Texas let energy demand grow without conservation incentives. As a result, during that 7-year period, energy use in Texas grew by 1.7 percent, compared with growth of 1.2 percent for the United States as a whole, while California actually reduced energy consumption by 0.5 percent. California encouraged conservation and cogeneration in industry, set mandatory building and appliance efficiency standards, and implemented major utility conservation programs, while Texas left energy use entirely to the marketplace. California's conservation initiatives clearly resulted in substantial economic and social savings. . . .

Conclusion

Most nations have only begun to exploit the opportunities for improved energy efficiency and the potential of renewable energy sources. In an interdependent world, there are good economic, political, and environmental reasons for the industrial countries, which use most of the world's energy supplies, to take the lead in improving efficiency and developing renewables, and to help Third World countries acquire the means to develop their energy resources in ways that will ensure efficient energy use on a sustainable basis. National and international development agencies, business and industry, and nongovernmental organizations all have vital roles to play in achieving these goals.

The U.S. Government Should Promote Energy Efficiency Programs

Eric Hirst

About the Author: *Eric Hirst is a corporate fellow in the Energy Division at Oak Ridge National Laboratory in Oak Ridge, Tennessee. He has written several articles on energy conservation.*

Improving the efficiency of energy use in the United States is the least expensive and most effective way to address simultaneously several national issues. Improving efficiency saves money for consumers, increases economic productivity and international competitiveness, enhances national security by lowering oil imports, reduces the adverse environmental effects of energy production (including acid rain and global warming), and responds to public interest in energy efficiency. Although much has been accomplished since the mid-1970s, there are still many opportunities to improve efficiency, and the federal government can play a major role in ensuring that those opportunities are not wasted.

A report commissioned for the U.S. Department of Energy's (DOE) National Energy Strategy described three possible scenarios for future energy use in the United States. One scenario, called "where we are headed," examines future trends in energy use in the residential, commercial, industrial, and transportation sectors of the United States from 1988 to 2010. According to this scenario, in which current government policies and programs remain in effect, annual U.S. energy consumption is projected to increase from the current level of about 81 quadrillion British thermal units (Btu) to 102 quadrillion Btu in 2010.

Another scenario, called "cost-effective efficiency," examines what could be achieved if market barriers to increased energy efficiency were overcome. These barriers prevent consumers in all sectors of the economy from adopting what would otherwise be economically attractive investments in energy-efficiency improvements. This scenario projects an annual consumption of only 88 quadrillion Btu in 2010 and reflects what could happen if government policies changed.

In comparison with these two projections, the consumption projected under a "frozen efficiency" scenario—115 quadrillion Btu in 2010—suggests that the United States might achieve only half of the cost-effective potential for improving energy efficiency if no more efficient technologies are developed and applied. Thus, new government policies and programs could double the energy savings expected to be achieved during the next 20 years.

"New government policies and programs could double the energy savings expected to be achieved during the next 20 years."

Unfortunately, many barriers stand in the way of achieving cost-effective efficiency improvements. These barriers fall into two groups: structural and behavioral. Structural barriers, such as fluctuations in fuel prices, government fiscal and regulatory policies, and infrastructure limitations on supply, result from the actions of manufacturers, builders, wholesalers, fuel suppliers, financial institutions, and government agencies and are primarily beyond the control of individual energy users. Behavioral barriers, on

Eric Hirst, "Boosting U.S. Energy Efficiency Through Federal Action," *Environment*, vol. 33, no. 2, pp. 7-10, 33-35, March 1991. Reprinted with permission of the Helen Dwight Reid Educational Foundation. Published by Heldref Publications, 4000 Albemarle St. NW, Washington, DC 20016.

the other hand, arise from end users' attitudes toward energy efficiency; the perceived risks of energy-efficiency investments; misinformation; and the lack of efficiency incentives.

Recent trends in U.S. energy use show that the gap between potential and actual efficiency improvements is widening. After more than a decade of steady improvements, the ratio of U.S. energy use to gross national product (E/GNP) stopped declining in 1986 and has remained virtually unchanged since then. Specifically, E/GNP declined by 26 percent between 1973 and 1986, but over the following three years, the ratio declined only 2 percent. Also, U.S. dependence on oil imports has increased from 32 percent of total oil use in 1985 to 47 percent in 1989.

Increasing the Federal Role

There are six primary reasons for increasing government action to improve U.S. energy efficiency. First, energy efficiency is a proven resource. As the evidence from 1973 to 1986 shows, the efficiency levels of homes, commercial buildings, industrial processes, and transportation systems have all improved dramatically. However, large opportunities remain for further energy-efficiency gains.

Second, U.S. retail prices of fuels and electricity do not reflect many of the problems associated with the production, transportation, and conversion of fuels. For instance, the price of a gallon of gasoline does not cover the costs of air and water pollution, oil spills, and dependence on foreign oil supplies.

Third, market imperfections keep consumers from making what would otherwise be economically rational choices. Fuel prices often do not reflect the marginal economic costs of providing the energy. Utilities, for example, base electricity prices on the average cost of providing the power, while the costs of constructing and operating new power plants, which are not passed on to consumers until after construction is completed, are usually much higher than the operating and capital costs of existing plants.

Another example of market imperfections is the inaccessibility of efficiency information. Consumers often find it difficult to get information on energy efficiency, and, even when information is available, it is often presented too technically for most consumers. Moreover, many consumers are unable to affect the efficiency of the energy-using systems they operate. For example, tenants in a large office building have little influence over the efficiency of the lighting, heating, and air-conditioning systems. And builders, not home buyers, usually select the appliances and heating and air conditioning systems for new homes. Not surprisingly, builders focus more on initial cost than on operating costs.

Thus, market imperfections can lead to decisions that weigh initial capital costs more heavily than reductions in long-term operating costs. These decisions are implicitly based on very high consumer discount rates, which reflect the value of money over time. The difference between discount rates used in making energy-efficiency investments (20 percent to 100 percent) and those used in making energy-production investments (about 10 percent) is called the pay-back gap. Because of this pay-back gap, investments in energy production are much more likely to be made than are investments in energy efficiency.

"Public opinion supports greater efforts to improve energy efficiency."

Government programs that overcome market barriers implicitly reduce consumer discount rates and lead to the selection of higher-efficiency systems. Increasing the energy efficiency of automobiles and residential space-heating equipment does not increase total purchase and operating costs. Therefore, government programs can provide large social benefits at no cost to individual consumers.

Fourth, energy-efficiency programs provide insurance: Even if concerns about global warming, acid rain, economic competitiveness, and the balance-of-payments are wrong, it still makes sense to improve efficiency. Increasing efficiency produces economic benefits for the nation because money saved through reductions in fuel costs is available for industrial investment and consumer purchases. For instance, the U.S. government spent $16 billion during the past decade to fill the Strategic Petroleum Reserve. Investments in energy efficiency would provide comparable insurance against increases in the price or restrictions in the supply of oil. Energy efficiency also adds diversity and flexibility to the nation's energy system. Unlike most supply projects (e.g., development of a major oil field or construction of a large power plant), efficiency improvements are small and use a variety of technologies. Even if a particular initiative proves ineffective, the loss will be small and will be more than offset by the many other successful energy-saving initiatives.

Energy Production

Fifth, federal fiscal policy historically has favored energy production over energy efficiency. In 1984, the federal government gave more than $40 billion to energy-supply industries, primarily through tax breaks. In comparison, energy efficiency received only 6 percent of the U.S. government's roughly $2 billion energy research and development (R&D) budget in 1988.

Finally, public opinion supports greater efforts to improve energy efficiency. When asked to choose the most important energy resource on which the government should spend more money, two-thirds of the respondents in a 1988 survey chose "renewables" or improved energy efficiency.

The U.S. Department of Energy has begun to recognize its responsibility to improve domestic energy efficiency. According to DOE's interim report on the National Energy Strategy,

> the loudest single message [from the testimony prepared for DOE's public hearings] was to increase energy efficiency in every sector of energy use. Energy efficiency was seen as a way to reduce pollution, reduce dependence on imports, and reduce the cost of energy.

There are a variety of ways by which DOE could improve U.S. energy efficiency: establishing a national energy policy with a strong commitment to efficiency; expanding the Federal Energy Management Program to implement efficiency improvements in federal facilities; providing grants and technical assistance to state and local governments; funding energy-efficiency research and development; helping state public utility commissions and utilities to promote regulatory reform and energy-efficiency programs; and adopting energy-efficiency standards.

"The average energy savings per retrofit home could be increased by as much as 50 percent."

In January 1990, DOE announced a new set of initiatives that should strengthen the role of the government in promoting efficient energy use. The proposal to "relamp" federal facilities with energy-efficient lighting fixtures, for example, will demonstrate the performance of new lighting technologies, encourage owners of commercial buildings to adopt their own measures, and save money for U.S. taxpayers. The companion initiative, designed to help the private sector relamp commercial buildings, could save businesses $5 billion per year in electricity bills. Similarly, the initiative to expand DOE's Integrated Resource Planning Program will benefit electric and gas utilities and their customers. In this program, DOE will work with utilities and regulators to improve resource planning, develop better data (especially on conservation and load-management programs), and create new computer models to analyze alternative resource strategies.

Although these new initiatives are encouraging, the DOE commitment to improve energy efficiency can be measured best by its deeds and dollars. Funding for DOE conservation R&D in 1990 was $194 million, and funding for DOE state and local assistance was $217 million, making the total DOE conservation budget $411 million. DOE's budget request for 1991 called for a 6 percent decrease in funding for R&D and an 86 percent decrease for state and local programs. Thus, the 1991 funding request of $213 million is barely half of the 1990 budget. It is difficult to reconcile DOE's enthusiastic words about energy efficiency with a budget request that proposes to cut the programs in half. . . .

Beginning in the early 1970s, most states established state energy offices to design and manage a variety of programs, often funded with DOE grants. The programs include demonstrations of, dissemination of information on, and financial aid for installation of energy-efficient systems.

The largest DOE grant program is the Weatherization Assistance Program, which provides financial assistance for retrofits to reduce heating and air-conditioning costs for low-income households. Since 1973, about 4 million homes have been retrofit, accounting for 19 percent of the 22 million homes eligible nationwide. The program annually receives $400 million from various sources (less than half of which is the DOE grant) and retrofits about 250,000 homes each year. The most effective conservation measures include attic insulation, wall insulation, furnace tune-up, and furnace replacement. Fuel savings per home range from 15 to 25 percent.

Current Funding

At current funding levels, it will take several decades to retrofit all of the eligible low-income homes in the United States. Therefore, it is important to expand such funding, develop new ways to increase the program's productivity, and work more closely with other organizations that deal with low-income households. Many utilities now run programs that are aimed at their low-income customers, and improving coordination between the federal and utility programs could increase program coverage and benefits. Additional research, such as the development and testing of improved energy audits to identify more accurately the costs and benefits of a wider array of retrofit measures, could increase energy savings and cost effectiveness. The average energy savings per retrofit home could be increased by as much as 50 percent.

"The success of government standards . . . shows that regulation can be a powerful tool in controlling energy use."

Since 1979, DOE's Institutional Conservation Program, which provides matching grants to retrofit schools and hospitals, has retrofit 64,000 institutional buildings at a cost of $800 million. Because funding is limited, states have developed innovative ways to expand the program. For instance, in 1985, Iowa created a Facilities Improvement Corporation to retrofit public and nonprofit buildings statewide. The corporation issues tax-exempt bonds to pay for the implementation of energy-efficiency measures with a pay-back period of six years or fewer. The corporation leases the efficiency measures to the building owners, but rental payments are designed to be less than the savings in energy costs. The money repaid to the corporation is then recycled to finance additional retrofits. The goal of the Iowa program is to complete energy improvements in all state buildings, schools, hospitals, local-government buildings, and nonprofit institutions by 1995. As of June 1990, engineering analyses had been completed for 345 school buildings and more than $3 million in improvements had been started.

Another state agency established to offset limited DOE fundings, the Arkansas Energy Office, makes detailed engineering audits of schools and hospitals to ensure that DOE funds are used as cost-effectively as possible. Arkansas permits only registered professional engineers approved by a state board to conduct these audits. The state also developed a computerized energy audit that costs one-third of what a typical engineering analysis costs per square foot of floor area because it is aimed specifically at public school buildings in Arkansas.

Another state program, the Energy Advisory Service to Industry, was set up by the New York State Energy Office in 1979. Focused on industrial firms with fewer than 400 employees, this service offers on-site energy surveys conducted by trained professionals at no cost to the manufacturer. Each survey includes computer calculations of the expected costs and energy savings of various energy-saving measures. The manufacturer receives a report detailing recommendations on lighting, motors, heating, air conditioning, boilers, and other energy-using systems in the factory. The report also suggests sources of financial assistance, including low-interest loans from the state, for installation of efficiency improvements. Since 1979, almost 10,000 surveys have been conducted, and the energy-efficiency measures installed because of the surveys are saving New York manufacturers about $70 million per year in lower fuel and electricity bills.

Energy Efficiency Standards

Under the 1975 U.S. Energy Policy and Conservation Act, automobile manufacturers were required to make improvements in new-car fuel economy, or mileage. Largely because of this federal standard, average new-car mileage in the United States almost doubled from 17 miles per gallon (mpg) in 1976 to 28 mpg in 1988, while new-car performance and amenities remained largely unchanged. However, the average fuel economy of new cars dropped in 1989 and again in 1990 because of declines in gasoline prices and federal administration opposition to the standard. New legislation that raises the fuel economy standard to 40 mpg by 2000 would cut gasoline use by 20 percent in the year 2010. Improved fuel economy could reduce carbon dioxide emissions and cut U.S. dependence on foreign oil.

By increasing the price of inefficient new cars and offering a rebate to buyers of fuel-efficient models, the United States could increase fuel economy. Such a tax/rebate system could be revenue-neutral for the auto industry because the fees on inefficient cars would be offset by the rebates for fuel-efficient cars. This system in conjunction with tighter government standards would push manufacturers to develop new technologies and encourage consumers to demand more efficient cars.

"New legislation that raises the fuel economy standard to 40 mpg by 2000 would cut gasoline use."

After Congress passed the National Appliance Energy Conservation Act in 1987, DOE developed standards that cover the major energy-using household appliances, including refrigerators, freezers, air conditioners, water heaters, and furnaces. The 1990 standards have increased the purchase price for many new appliances but cut their operating costs. The pay-back periods range from less than one year to almost eight years, with an average of about two years. Thus, the standards require improvements that are very cost effective and are likely to save consumers roughly $40 billion by 2015.

The typical pre-1990, 18-cubic-foot refrigerator/freezer consumed 1,065 kilowatt-hours (kWh) per year. The 1990 standard reduced that amount to 960 kWh per year, and the 1993 standard should cut energy consumption another 28 percent, to 690 kWh per year. Nationwide, the savings from refrigerators and freezers will amount to almost 4 percent of projected residential electricity use in 2015. In addition, the reduced energy consumption of the new appliances will eliminate the need to build 15 large (500-megawatt) power plants.

The success of government standards for new cars and appliances shows that regulation can be a powerful tool in controlling energy use. National standards can overcome the barriers to energy efficiency, save money (on electric and fuel bills, for instance), and provide many other social benefits. The possibilities for developing standards for other energy-using devices should be closely examined. Congress recently authorized standards for fluorescent lamp ballasts, but the costs and benefits of developing national standards for all lamps and fixtures should be examined (and for motors, as well).

The Need for Action

DOE should work closely with builders and state and local governments to develop, implement, and enforce energy-efficiency codes for new construction. When instituted, construction standards have been very effective in improving the energy efficiency of new buildings—especially on the West Coast. In Washington State, thermal-performance codes for new residential buildings have been revised three times since 1977 and receive strong support from the state energy office, other government agencies, electric and gas utilities, environmental groups, the association of county and city governments, and even builders. The Washington State Energy Office, which has played important roles in the development, passage, and implementation of these codes, conducts experimental and analytical projects that provide data on the energy savings, costs, and cost effectiveness of different levels of standards. These projects indicate that the 1990 standard will cut space-heating energy use by 60 percent for electrically heated homes and by 40 percent for homes heated with gas, relative to the original 1977 standard. In addition, the office trains local code officials, thus helping to ensure that builders comply with the code.

At the request of Congress and with support from DOE, the National Research Council (NRC) has reviewed the nation's energy R&D programs. NRC recommends that DOE "allocate to energy programs on conservation and renewables an additional $300 million [per year] or about 20 percent of the civilian energy R&D budget." The report repeatedly emphasizes the large opportunities to improve efficiency and notes, for example, that energy use in residential and commercial buildings could be cut by more than 70 percent.

"Energy use in residential and commercial buildings could be cut by more than 70 percent."

Clearly, improving the efficiency of energy use in the United States could yield important benefits, including the direct savings from lower fuel bills, reduced emissions of greenhouse gases and other pollutants, increased independence from foreign oil, and improved economic productivity and international competitiveness. But the United States will enjoy these benefits only if the federal government actively promotes energy efficiency through increased research, expanded state and local programs, stronger efficiency standards, improved regulation of energy utilities, and increased technology transfer. Half of the possible cost-effective energy savings between now and 2010 will not be realized unless government policies and programs are changed.

The Bush Administration Must Improve Its National Energy Strategy

Will Nixon

About the Author: *Will Nixon is the associate editor of* E Magazine, *a bimonthly environmental magazine.*

Just when the peace movement had run low on placard slogans—"No Blood For Oil" having lost some of its octane as the Gulf War wound down—George Bush announced his long-awaited National Energy Strategy (NES) at lunchtime on February 20, 1991. Not since Jimmy Carter had a president decided that energy planning should involve more than checking the gas gauge before heading off on a long trip. Indeed, Carter had donned a cardigan sweater, turned the White House thermostat down to 68, and solemnly told the nation that his energy policy was "the moral equivalent of war." Cynics now quipped that George Bush had dropped the equivalency part. But the NES was much more than that. It admitted that the United States couldn't kick the foreign oil habit, which was the great dream of presidents in the 70s, but offered almost 100 ways to increase our own energy production. It recommended opening up Alaska's Arctic National Wildlife Refuge for oil and gas exploration, reviving the nuclear energy industry by eliminating some of those pesky public hearings, freeing the oil and gas industries from some cumbersome regulations, and burning more garbage for energy. As for encouraging us to use less energy to begin with, well, better luck next time. The NES virtually ignored better car mileage standards and mass transportation. It dropped tax breaks for renewable energy and federal lighting standards. And, perhaps most negligent of all, it never got around to the reason we need an energy policy in the first place—to stop the flood of carbon emissions into our atmosphere which cause global warming. "When people talk about the need for a national energy policy, they don't mean writing down the mistakes of the past 10 years, which is what the Bush Administration has done," said Alexandra Allen of Greenpeace.

The next morning, a handful of Greenpeacers went to the Senate offices where Admiral James Watkins, secretary of the Department of Energy (DOE) was due to appear before the Energy Committee. Senator Albert Gore of Tennessee had already promised a "battle royal" on Capitol Hill over Bush's energy plan. The Greenpeacers arrived at 7:30 AM, but the Washington media pack had already claimed the room. So they stood out front, holding a "Reality Check" for Bush and Watkins. It was made out for "Untold Billions" for "Oil Wars, Oil Spills, Global Warming and Nuclear Waste."

"[The NES] never got around to the reason we need an energy policy in the first place—to stop the flood of carbon emissions."

In the gray and white world of Washington, Greenpeace provides the most color, but they were hardly alone in their outrage at the president's energy plan. Only a year ago the DOE had suggested that its plan would be almost the reverse of what Bush finally announced. Deputy Secretary Henson Moore had said, "Energy efficiency and renewables are basically the cleanest, cheapest and safest means of meeting our nation's growing energy needs in the 1990s and beyond."

Will Nixon, "Energy for the Next Century," *E Magazine*, May/June 1991. Reprinted with permission.

"Based on comments by the DOE last year," said Scott Denman of the Safe Energy Communications Council, "and based on George Bush's claims in his State of the Union address that his strategy would support conservation, efficiency and alternative fuels, what we've been given is a cruel hoax. It's not an energy strategy, it's an energy tragedy."

"Based on George Bush's claims . . . that his strategy would support conservation, efficiency and alternative fuels, what we've been given is a cruel hoax."

Out in San Francisco, Chris Calwell of the Natural Resources Defense Council (NRDC) said, "The themes that emerge are simple and familiar: drill more, nuke more, pay more, save less." His group released an analysis entitled "Looking for Oil in All the Wrong Places," with some math to show just how wrongheaded the NES was. If we drilled the Arctic Refuge and the Outer Continental Shelf, as the NES proposed, we might produce a little more than six billion barrels of oil or its equivalent in natural gas, and these areas would be drained dry by the year 2020. But if we used our renewable energy supplies—the sun, wind, rivers that flow into hydroelectric dams, and plants and trees that can be turned into biomass fuels—and we made our world far more energy efficient, we would have the equivalent of 66 billion barrels of oil. As Robert Watson, the author of this report, says, "We *will* have a post-petroleum society. The question is whether it will be like Mad Max or Ecotopia."

Expert Testimony

The NES hadn't always looked so grim. For more than a year, the DOE had held hearings across the country, patiently listening to an army of experts who hadn't found a sympathetic ear in the executive branch since Ronald Reagan took office and put the DOE under the sleepy care of dentist James Edwards. Admiral Watkins had made his name in Washington, DC by taking over Reagan's troubled AIDS commission and producing a good report. "He showed he has a good ear. He's receptive to contrary points of view," says Scott Denman, who spoke at two hearings late in 1989. "He heard from every energy expert in the country."

These experts had a remarkable story to tell: the United States, with what scrambled energy policies we do have, already saves an estimated $160 billion a year in energy costs compared with 1973. "The really interesting thing is that total energy consumption was virtually identical in 1973 and 1986," says John Morrill of the American Council for an Energy-Efficient Economy. "And yet our GNP [gross national product] grew by 40 percent during those years."

Until the 70s, conventional wisdom had insisted that energy growth was synonymous with economic growth, but changes wrought by the oil shocks of 1973 and 1979 tied that thinking in a knot. The Corporate Automobile Fuel Efficiency (CAFE) standards passed in 1975 pushed cars from 13 miles per gallon to 27.5 in 1986, and now save us five million barrels of oil a day. Utilities dropped oil for cheaper coal. Homes were weatherized. Some states took action, such as California, which required 1980 refrigerators to use 20 percent less electricity than the 1975 models, causing an entire industry to change. (By the mid-80s refrigerators, the biggest energy drains in most homes, had improved by 35 percent.) Not until 1985 when President Reagan rolled back the CAFE standards and oil prices dropped, did our energy savings stall.

One after another people appeared before the DOE with ideas on how to get us back on the savings track. After all, we still have a ways to go—Japan and Germany use half as much energy per dollar of economic output as we do. "The message from elected officials, citizens' groups and energy related businesses was conservation, conservation, conservation," Denman says.

The DOE listened. "Admiral Watkins looked

at the issue, and he began to say the one thing that's loud and clear is that improved conservation and energy efficiency are where the most gains can be made and the government can do a lot more than it has," said Christopher Flavin of Worldwatch who testified at the first hearing in August 1989. In April 1990 the DOE released an interim compendium of what it had heard so far which was dominated by conservation and energy efficiency. But then the report left the DOE for the White House.

Some States Conserve

"President Bush knows as much about energy as anyone we've ever had in the White House," says one DOE official. What Bush knows, though, is producing energy, not saving it. His home state, Texas, is the worst energy hog in the country. It can't even fill up its own gas tanks, importing 48 thousand barrels of oil in 1989 on top of 736 thousand barrels from its own wells. "Texas has 42 percent fewer citizens than does California," reports a study by the consumer advocacy group, Public Citizen, "but it uses 37 percent more energy." California has a state energy department and Texas doesn't, which just shows what a little planning can do.

"The message from elected officials, citizens' groups and energy related businesses was conservation, conservation, conservation."

"The implicit assumption of the NES is that the market runs at top efficiency," says Michael Brower of the Union of Concerned Scientists (UCS). "Any conservation or renewable energy we have is the amount we should be getting because we have a free market. Our point is that the market is not really free. There are all sorts of incentives for fossil fuels which, to be fair, should be eliminated or applied to energy efficiency and renewables. And the free market doesn't take into account the cost to the environment." This spring the UCS, NRDC, and the Boston-based nonprofit environmental research group, the Tellus Institute, released an alternative national energy strategy to show what we can do by taking conservation and efficiency seriously. "Nobody has ever made this serious an effort to meet the skeptics on their own ground," Brower says.

More than a dozen states now give serious consideration to energy efficiency and the environmental costs of fossil fuels when planning out their needs. Rather than simply worrying about energy supply, they consider demand, and often find that the cost of reducing the demand is a lot cheaper than increasing supply. This new approach, labeled "demand side" or "least cost" planning, now dominates on the West Coast and in New England where states have to import fossil fuels. States that produce coal and oil, though, tend to stick to the "supply side" view.

"If every state took the least cost approach there would be no reason to build any new power plants—coal or nuclear—in this decade," says Michael Totten of the International Institute for Energy Conservation. And if we took this approach in all of our energy decisions, we could be using seven percent less energy by the year 2000, instead of 34 percent more, as we would on our present course. In the next century, after we've captured these savings, renewable energy sources such as windmills, photovoltaics that turn sunlight into electricity, and turbines that run on biomass fuels could be competitive with fossil fuels.

But the NES took the supply side view. And many observers laid the blame on John Sununu more than on George Bush. "Sununu has always been hostile to demand side measures," Robert Watson says. "We both think the other's solution is a drop in the bucket. He dismisses conservation out of hand as being irrelevant. We dismiss nuclear power as largely being irrelevant. But I think we have the better basis for dismissing it."

Should the U.S. Expand Its National Energy Strategy?

No: The U.S. Does Not Need to Expand Its National Energy Strategy

A National Energy Strategy Is Unnecessary
A National Energy Strategy Would Harm the U.S.
The Bush Administration's National Energy Strategy Will Work

A National Energy Strategy Is Unnecessary

Virginia I. Postrel

About the Author: *Virginia I. Postrel is the editor of* Reason, *a monthly magazine published by the Reason Foundation, which is devoted to the principles of capitalism and individual liberty.*

Ever since 1973, Americans have been crazy about energy. The trauma of that year's gas lines and heating-oil shortages—reinforced in 1979—has now outlived "the Vietnam syndrome." As U.S. troops return victorious from the Persian Gulf, the great off-stage agenda setter has determined that the next big issue will be energy policy. We oughta have one, everyone seems to think.

In part, the enthusiasm for an energy policy stems from the war itself. Had we had better policy, the purveyors of conventional wisdom declaim, we wouldn't have had a war in the Persian Gulf. "American blood soaked into the sands of the Middle East provides a tragic reminder to us all that this nation's political processes have failed to come to grips with our addiction to imported oil," writes John E. Petersen.

Ever accommodating, the Bush administration has sent a 214-page energy-policy proposal to Congress. Of course, nobody—at least no Democrat—likes it. Sen. Al Gore (D-Tenn.) called it "breathtakingly dumb." Like many critics of U.S. energy policy, he condemned U.S. energy use, employing the always-damning rhetoric of addiction. With its emphasis on expanding energy supplies, Gore declared, Bush's proposed policy is "kind of like somebody who is addicted to alcohol or drugs confronting the problem and deciding the solution is to get a lot more of it."

Virginia I. Postrel, "Fuelishness." Reprinted, with permission, from the May 1991 issue of *Reason* magazine. Copyright 1991 by the Reason Foundation, 2716 Ocean Park Blvd., Suite 1062, Santa Monica, CA 90405.

Alden Meyer of the Union of Concerned Scientists used similar language to express what many environmentalists want out of the American public: "As Alcoholics Anonymous has proven, admitting you have a problem is the first step on the road to recovery." First, we have to acknowledge that our dependence on energy is sick and dangerous. Then we have to kick the habit.

This is all nonsense, "breathtakingly dumb," to quote would-be President Gore. The United States has no more need for an "energy policy" than it has need for a shampoo policy or a socket-wrench policy. Fuel is just one good among many. And we are no more "addicted" to energy than we are to silicon or aluminum or anything else that goes into producing the goods and services traded in the U.S. economy. It is not evil to use oil to make things.

But what of the Persian Gulf? Weren't we willing to sacrifice blood for oil?

Well, no. Contrary to [Senator] Bob Dole, the Gulf War wasn't about o-i-l— at least not about keeping the oil flowing to U.S. consumers. The one thing oil-producing nations, including Iraq, will always want to do is sell oil.

"The United States has no more need for an 'energy policy' than it has need for a shampoo policy."

The problem with Iraq, aside from the general issue of aggression, wasn't that we'd lose access to oil. It was that controlling a lot of oil makes it possible to get very rich, which makes it possible to buy lots of weapons . . . which can lead to very unpleasant situations. The only energy policy that would have countered Iraq would have been the one that only the most honest and extreme greens advocate: not using any oil at all. And even that policy would have worked only if everyone else in the world went along with it.

When people bemoan "our dependence on foreign oil," whether they want to expand domestic supplies or cut consumption, they ignore a fundamental fact. Oil is oil. Its molecules don't come with little PUMPED IN SAUDI ARABIA labels. As MIT [Massachusetts Institute of Technology] energy economist M.A. Adelman writes in *Foreign Policy*, "The world oil market, like the world ocean, is one great pool. The price is the same at every border. Who exports the oil Americans consume is irrelevant."

The Flawed Efficiency Argument

Not everyone calling for a U.S. energy policy uses the hysterical language of addiction. Calmer voices speak instead of efficiency. We ought to conserve energy, they say, simply because it makes good sense. If a car getting 35 mpg can take you places as well as a car getting 20 mpg, you're a fool to buy the 20-mpg model. And the car companies are fools or villains to make it. Congress should pass a law to require cars to get better mileage.

The efficiency argument is appealing: Why not do more with less? But it ignores the big picture. Fixated on energy, its advocates tend to forget the rest of the economy. And they disregard the way real producers and consumers make decisions.

Consider the Sun Frost refrigerator. Touted by *Sierra* magazine's upbeat energy issue, the 16-cubic-foot model reportedly uses one-fourth to one-third as much energy as a standard refrigerator. The cost: a mere $2,400. An ordinary 16-cubic-foot fridge runs about $800—a third the cost of the nifty Sun Frost. Unless electricity rates are very high and interest rates are very low, you're better off saving the money up front.

Efficiency fans tend to make the Sun Frost mistake a lot. They forget that up-front investment, whether to retool a factory or buy a more-expensive refrigerator, matters a lot. A dollar spent today is worth more than a dollar saved in three years.

And it's not just money. The money *means* something. These prices aren't set by a czar in Washington (though when oil prices were set more or less that way we had those lines. . . .). Prices reflect the value producers and consumers put on goods, relative to other goods. If people do value good mileage—presumably, because gas prices are high—they'll pay more for cars that get 35 mpg.

But people may not be willing to pay extra for better mileage. Maybe they don't drive a lot. Maybe gas is cheap. Maybe they'd just rather spend the money on new clothes for the kids. Energy freaks want to make everyone else love gasoline the way they love gasoline, and, if not, they'll pass a law to get the same results.

"Energy freaks want to make everyone else love gasoline the way they love gasoline."

Even when people make energy-saving investments for completely voluntary reasons, those investments exact a price. When rising gas prices make airlines buy new, more efficient planes, they can't spend the money on employee training or improved safety equipment. And the capital they raise to pay for their new fleets can't go to fund startup companies, finance more-productive widget plants, or underwrite new housing.

Writes Adelman: "Throughout the 1970s and 1980s, investment in conservation measures, made in order to achieve lower energy consumption, inevitably diverted investment from other uses. Sluggish growth in productivity created sluggish growth in living standards." It's one thing to make such trade-offs because real energy prices are rising. It's another to force them on your fellow Americans because you think energy is the most important thing in the world.

A National Energy Strategy Would Harm the U.S.

Irwin Stelzer

About the Author: *Irwin Stelzer is a resident fellow at the American Enterprise Institute, a public policy think tank in Washington, D.C.*

Romanians do without electricity for large parts of the day. Poles and East Germans choke in the pollution from inefficient electric generating plants. The British public pays a 20 percent surcharge on its electric bills to pay for past nuclear sins of its then-nationalized power system. Russian oil production, suffering from years of government mismanagement, is in decline. China is unable to harness its vast resources of natural gas. Americans, meanwhile, have enjoyed ample supplies of relatively inexpensive oil and natural gas, and—so far—haven't had to worry too much about whether their lights will go on when they flip a switch.

The lesson seems obvious: market economies, even those with some regulated sectors, deliver adequate supplies of energy, when and where it is needed, at reasonable prices and with tolerable social costs; centrally directed, planned economies don't. Unfortunately, James Buchanan's observation that "the accumulation of empirical evidence must ultimately dispel romance" is not applicable to bureaucrats, even Republican bureaucrats, who inhabit the well-lit, air-conditioned offices of the White House and the Department of Energy.

So, in 1989 President George Bush directed the DOE (an agency Ronald Reagan and his vice president promised to abolish) "to lead an interagency effort to develop a National Energy Strategy . . . [to] serve as a blueprint for energy policy and program decisions." Pursuant to that directive, Secretary of Energy James Watkins "opened a dialogue with the American people," holding fifteen public hearings, receiving more than a thousand written submissions, and compiling a hearing record of some 12,000 pages. This resulted in a 230-page "Interim Report," which will form the basis for still more public comment, en route to a year-end "first edition of the National Energy Strategy." In the words of a triumphant DOE press release, "the report will provide a baseline for development and analysis of energy options."

The bipartisan applause for this effort is deafening. The General Accounting Office, not notable for its support of Republican initiatives, has only good things to say about this one: "We support the initiative to develop a national energy strategy and believe that such a strategy is sorely needed and long overdue." Rather than trust consumers and the marketplace, the GAO is eager for government to "propose initiatives for guiding future energy choices."...

"Government interventions in energy markets . . . lead to enormous inefficiencies."

Grizzled observers of past efforts to forge a national energy policy are taking tight hold of their wallets. If history teaches anything, it is that government interventions in energy markets create shortages (natural gas, gasoline), gluts (natural gas), raise costs (electricity), and lead to enormous inefficiencies.

This country has suffered through President Nixon's "Project Independence," launched in 1973 to enable us, in his words, "to meet our own energy needs without depending on any foreign sources." Since then, imported oil has risen from 35 percent to almost 50 percent of our supply. Nixon also treated us to a gasoline-

Irwin Stelzer, "National Energy Planning Redux." Reprinted with permission of the author from: *The Public Interest*, No. 101 (Fall 1990), pp. 43-54, © 1990 by National Affairs, Inc.

allocation scheme that led to long lines at all pumps except those on Capitol Hill that were reserved for congressmen. President Carter repeated that error by promulgating federal controls on thermostat settings, substituting sweaters for oil and gas heat and sweating for air conditioning. This he called the Moral Equivalent of War, a macho analogy to real war that was somewhat reduced in force by its unfortunate acronym—MEOW.

Government Policies

Presidents have not acted alone in this silliness. In 1978 Congress chipped in with the National Energy Conservation Policy Act. Section 682 of the act contains this "finding": "The Congress recognizes that bicycles are the most efficient means of transportation." The image of Tip O'Neill and Ted Kennedy pedaling up Capitol Hill to support this measure is amusing.

All of these efforts to manipulate the energy economy would be more amusing if they had not been so costly. After the first OPEC [Organization of Petroleum Exporting Countries] price shock, attempts to control oil prices cut billions off our GNP [gross national product] and contributed to the inflation that plagued the 1970s. Attempts to prevent the use of natural gas in utility boilers increased both electricity prices and pollution levels—all to conserve a fuel now in excess supply. Legislation to force electric companies to buy power from independent firms jacked up supply costs and contributed to an inefficient increase in generating capacity.

That President Bush should consider reinserting the government into the energy business should, on reflection, come as no surprise. As vice president he responded to a collapse in oil prices with a widely publicized appeal, made during a visit to Saudi Arabia, for stable (most observers took this to mean higher) oil prices. Fortunately, President Reagan shot down his lieutenant's effort to transfer large amounts of income from the vice president's oil-consuming Maine neighbors to his oil-producing Texas neighbors.

But President Bush's ardor for some sort of national energy strategy, rather than fluctuating, market-determined prices, remains uncooled. Like his predecessors—indeed, like most politicians here and abroad—he finds it difficult to leave energy pricing, and demand and supply decisions, to consumers and producers interacting in a free market.

The most obvious reason for continued government attention to the energy economy is that powerful pressure groups are affected by what goes on in those markets. Tales of oil barons and their more modern corporate counterparts lobbying for government favors dot American history books. From the days of Teapot Dome, and earlier, the oil industry has had an intense interest in the leasing of oil lands. It also has had a huge stake in import policy, at times winning quotas or tariff protection that would turn Lee Iacocca green with envy.

"The imposition of price controls in the 1970s encouraged excessive consumption."

In short, government action (or inaction) in energy markets means money. Higher oil or natural-gas prices send a huge flow of dollars from New England and the Midwest southward, just as surely as will the taxpayer bailout of the savings-and-loan depositors. Lower or controlled prices reverse the regional flow, conferring windfalls on northern consumers, and devastating the economy of Texas—the home state at various times of presidents, Senate and House leaders, and key cabinet officers. The *Economist* estimates that every one-dollar fall in the price of oil costs Texas 25,000 jobs. The stakes are simply too high for politicians to adopt a consistent hands-off attitude toward energy markets.

These financial motives are so transparent, and so easily understood, that actions taken in pursuit of them only occasionally cause problems. When oil companies bribe federal officials (Teapot Dome) or attempt to buy senators (natural gas decontrol), or when oil-industry opponents raise a cry about "obscene profits," the issues are on the table for all to see.

"It is one thing to pay more for gasoline, another not to be able to get any because some bureaucrat allocated too much to the wrong place."

This is not the case with those who use energy policy as an instrument for implementing their broad social agenda. In the late 1970s we were treated to the Club of Rome and similar groups' fears that the world was running out of resources and that stringent efforts to curb consumption were essential. This view had its roots in the sumptuary mentality of the Middle Ages, derived in turn from ancient Greek and Roman statutes limiting expenditures on funerals, per-capita entertainment, and dress. Malthus resurrected this nervousness about over-consumption in a mixture of economics and moral judgments that found its way into President Carter's piety: "[T]oo many of us now tend to worship self-indulgence and consumption. . . . [P]iling up material goods cannot fill the emptiness of lives which have no confidence or purpose." In short, small is beautiful, less is more. It was a short jump from that view to establishing a national energy policy that was "the moral equivalent of war," to be won by establishing "a new conservation ethic." For if one believes that Americans are overly self-indulgent consumers of material goods, energy policy is the perfect tool to exorcise that devil. Car pools instead of stereophonic privacy when driving to work; the equality of mass transit, enforced by restrictions on gasoline and vehicle use; summer thermostats at 78°, winter settings at 65°; frosty refrigerators and manually cleaned ovens, substituting manpower for electric power: all have been ingredients of various national energy plans. All, if effectively implemented (often an impossibility), save energy and satisfy the yearnings of the Malthuses, Carters, and their ilk for a return to a simpler, more austere, and therefore more virtuous life.

Such a change, whatever one thinks of it, in what we have come to call "life style" is only tangentially related to energy policy, which is merely the lever used to pry Americans from their attachment to material things. It should come as no surprise that groups that worry about Americans' attachment to material goods oppose nuclear power as too dangerous, coal as too dirty, and oil as environmentally unacceptable because of the effect of offshore drilling and, lately, oil spills. Their preference is conservation— reduced demand rather than increased supply.

Oil and Our National Security

Not all proponents of a national energy policy share this minimalist view of the good life, of course. Others worry about national security, which they believe is threatened by an excessive reliance on imported oil. With America now dependent on such supplies for almost 50 percent of its oil needs, those concerns are not irrational. After all, the oil embargo and the oil price "shocks" of the 1970s did contribute to the stagflation of the late 1970s, costing us dearly in income and jobs. And the more recent Middle East crisis may well produce more of the same. This can be avoided, interventionists argue, by subsidizing domestic production, setting import quotas, or mandating reductions in gasoline consumption.

There are two responses to this argument for a national energy policy. The first is that the 1990s are not the 1970s. We have learned, for one thing, that price controls are not an efficient response to supply interruptions or price spikes. The imposition of price controls in the 1970s encouraged excessive consumption in response to the artificially lower prices and discouraged the development of domestic supplies.

We have also learned the modern value of Joseph's Biblical advice to the Pharaoh about lean years following fat ones. Our strategic reserve of oil now totals 600 million barrels, the equivalent of eighty-one days' supply. In addition, we have diversified our sources of crude-oil supplies: 70 percent of our imports come from Venezuela, Mexico, Canada, and other nations outside of Arab OPEC.

The second response to those who see a national energy policy as necessary to national security is more complicated, since it cuts across several of the energy industries. It stresses the need to adopt a program that makes markets work better, rather than one that substitutes the heavy hand of government for the invisible hand of Adam Smith.

Economists have long recognized that markets efficiently allocate resources only so long as the prices attached to the products in question correctly reflect the costs of producing those products. Those costs include the labor and capital employed in the production process, and the cost of what on the surface appear to be free goods, clean air and water being the most notable. It is now widely accepted that many production processes, especially those involving the burning of fossil fuels, consume environmental resources. To the extent that the cost of those resources is not reflected in the prices of those fuels, consumers receive an erroneously low price signal, causing them to overconsume. That much seems clear to all parties to the dispute over energy policy, from Governor John Sununu to the greenest environmentalist. Their squabble is over how much these environmental resources cost, and how best to force producers to reflect those costs in prices.

"Energy forecasting . . . is a mixture of analysis and self-serving numbers manipulation."

Not so obvious, except when some crisis bursts upon us, is the fact that the importation of oil also imposes costs that are not reflected in the price of crude oil. For one thing, the impact of imports on our rate of inflation, balance of payments, and national defense, combined with the effect of our incremental consumption on world oil prices, means that current oil prices do not reflect all the externalities associated with our level of oil imports: even the cartel price is below the true marginal cost of that oil to us. This means that America consumes more imported oil than is economically optimal.

Consumption Levels

To achieve an optimal level of consumption, the price of imported oil must be raised high enough to reflect its true marginal cost to society. This can most efficiently be accomplished by a tariff of about $10 per barrel, according to an estimate by Harvard's William Hogan. Such a device would, of course, adversely affect domestic oil consumers. But it would do so only because those consumers are not now paying the full cost that their oil consumption imposes on society. In short, a properly conceived tariff would end their subsidy. And it would sop up economic rents otherwise available to the OPEC cartel and its fellow travellers, while at the same time providing a more efficient incentive to the development of new technologies than would the subsidization of the "promising" technologies so beloved of environmentalists (solar, wind), construction engineers (synthetic fuels), and farmers (grain-based motor fuel).

Note that imposition of such a tariff is entirely consistent with leaving the energy business to the market. It merely gets the price signals right, and then lets consumers decide how much imported oil to use, and domestic energy producers how much oil (or alternative fuels) to produce. The alternative of doing nothing encourages overconsumption of deceptively low-priced imported oil. And the alternative of quotas involves the government in deciding just who gets how much of the bargain-basement petroleum.

The failure of past allocation systems, many devised in response to previous Middle East eruptions, is proof enough that the most recent (Iraqi) threat is no basis for an interventionist national energy policy. Saddam may get prices up, largely because we failed to do so first with a tariff. But higher prices produce far fewer dislocations than a government rationing scheme—it is one thing to pay more for gasoline, another not to be able to get any because some bureaucrat allocated too much to the wrong place. And a policy of getting prices right, by internalizing the environmental costs of fossil-fuel use and reflecting the economic and security costs of imported oil in its price, would have other salutary effects. It would increase the relative attractiveness of conservation techniques—with energy prices at higher, economically efficient levels, it would pay for more people to insulate their homes, drive more slowly, and turn off the lights when leaving home—without a web of government rules limiting energy consumption. And it would encourage the reappraisal of nuclear power now being whispered in policy-making circles, by changing the relative costs of fossil fuels and nuclear-generated power.

The American nuclear program is now moribund. Although some 112 licensed and operating nuclear reactors now provide about 20 percent of the nation's electricity, no new plants have been ordered since 1978, and several have been cancelled. The reasons for this state of affairs include inept management of, and huge cost overruns on, nuclear projects; changing federal safety regulations; state rate regulations that levy *ex post facto* penalties on companies investing in plants that hindsight proved not to be needed, just yet; an inability to devise a politically acceptable means of disposing of nuclear waste; media that consider Jane Fonda an authoritative source of scientific information; politicians who pander to local hysteria by closing down plants that are completed and ready to produce needed energy; and, until recently, a slowdown in the growth of demand for electrical energy.

The Nuclear Program

Whether or not the nuclear program in America will recover—with newly designed, smaller, and safer plants having a role in the next century—is difficult to determine. Experience has taught that it is foolish to rely on the optimistically low cost estimates of committed technologists, or on the counterclaims of conservationists who specialize in very high cost estimates. Both proponents and opponents of nuclear power regularly fail to concede the obvious: predicting the costs of nuclear power and comparing them with those of oil, gas, and coal plants is extraordinarily difficult. What Oxford's George Yarrow calls "the balance of advantage between the alternatives" depends on the course of coal, oil, and gas prices well into the next century, the long-term level of interest rates, and the course of environmental and safety regulation—variables that no responsible economist can forecast with an easy conscience .

"The American nuclear program is now moribund."

Fortunately, there is no need for policy makers to adjudicate these disputes over cost projections: that's what the market is for. For the past decade the market has been saying that it will not commit capital to the construction of nuclear plants, given the prospects for demand growth, the costs of constructing the plants, and the regulatory environment in which the electric-utility industry operates. Ordinarily, those who support market-determined solutions to energy problems would applaud the decisions to cancel plants, and the refusal of the industry to start new ones. But the world of nuclear power is no ordinary one.

For the current sorry state of our nuclear industry is due only in part to the low demand and high cost projections that entrepreneurs have faced. It is due equally to the asymmetry of risk created by state regulators. If a nuclear plant is built on time and within the projected cost, the utility is permitted to earn a modest, so-called "reasonable" return on that investment. If the power produced proves to be cheaper than that from fossil-fuel plants, the benefits are passed on to consumers in the form of lower electric rates. But if the plant significantly overruns its cost projections, or is late, or proves to add to excess capacity for some time (these plants take a decade to build, and often are not "needed" at the moment of completion), the investor is likely to face a regulator who will deny him a return on, and possibly a return of, the billions of dollars of capital committed to the plant. Unless this asymmetry is eliminated, markets cannot perform their function of determining whether it is appropriate for investors to risk their capital on the construction of nuclear plants to meet future power needs. If we are ever to find out who is right—those who think that nuclear power has a cost-effective role to play in our energy future, or those who argue that it is wildly expensive, and that small gas-fired plants and conservation are more efficient—this asymmetry must be eliminated. We don't need nuclear subsidies; we do need a system of regulation that permits markets to work—that is, to reward entrepreneurs who guess right and to penalize those who guess wrong.

Forecasting Energy Demand

Until recently, resolution of this issue was more important to the private investors who had committed billions to nuclear plants than it was to the broader national interest. With substantial excess generating capacity available in most parts of the country, few large power plants of any sort were needed. But that circumstance may well be changing. Economists at the Federal Reserve Bank of Boston have reported what New Englanders know: utilities in the region have had to invoke "emergency operating procedures in order to cope with the region's peak demands for electricity." New York utilities may, but will not certainly, be able to meet projected demands later in this decade. Wisconsin utilities and regulators are reported in the trade press to be scrambling to figure out how to supply the power their customers will need in a few years. Florida utilities resorted to rolling blackouts when severe cold weather hit the state in the winter of 1989-90. "We . . . will need to build a lot of generating capacity over the next ten years," concludes Federal Energy Regulatory Commission chairman Martin Allday. Now seems to be the time when the market should be freed to appraise the nuclear option.

"The joke in the trade is that energy forecasters were invented to make economic forecasters look good."

This conclusion seems unassailable even if we recognize that forecasts of energy demand are notorious for their unreliability; the joke in the trade is that energy forecasters were invented to make economic forecasters look good. The reasons for past failures to forecast the demand for energy with any accuracy are partly technical and partly institutional. On the technical side, such forecasts rely on elaborate econometric models, which depend heavily not only on data relating to the energy industry (prices, quantities demanded, etc.), but also on forecasts of GNP, one of the driving forces behind energy demand. As a consequence, any error in forecasting the general level of economic activity—and current debates over whether we are headed into a recession or are to be blessed with a renewed spate of growth suggest how difficult economic forecasting is—has a magnified effect on energy-demand forecasts.

Equally important is the fact that energy forecasting, as it has been practiced, is a mixture of analysis and self-serving numbers manipulation. Since the financial community now tends to frown upon utilities with large construction programs—the fear being that regulators will not permit an adequate return on newly invested capital—the pressure on all utilities is to forecast zero or low growth, so that analysts will be persuaded that the company will not be raising and investing new capital in the industry. On the other hand, a new generation of would-be energy czars prefers a high-growth scenario, since it desperately needs an impending shortage in order to support its arguments for government intervention in one form or another. After all, a crash program to build subsidized nuclear plants, or to require people to use less energy, or to require more fuel-efficient cars, makes no sense unless there is a projected shortage of energy.

The good news is that whereas central planners need consensus forecasts, market economies don't. Market-driven economies are ecumenical enough to include those who predict oversupply, and those who see an opportunity for profit in building new plants. The latter risk their capital, in the hope of profit if they guess right, in the certainty of losses if they guess wrong.

Unfortunately, that description of a functioning capitalist response to consumer needs doesn't quite fit the existent energy market. And both the private and public sectors are to blame.

On the private side, large elements of monopoly power operate to prevent supply from efficiently meeting demand. In the electric-power industry some companies can use their control of transmission capacity to prevent competitors from getting cheaper power to market, and railroads use their control of rights-of-way to prevent coal-slurry pipelines from competing effectively with rail carriers .

On the public side we have statutes that require utilities to buy power from inefficient suppliers, regulations that have been used to confiscate shareholders' investments in power plants, and subsidies for various forms of energy that appeal to one or another set of legislators.

Given this leaky fountain pen, the federal energy bureaucracy is preparing to don a pair of rubber gloves—the National Energy Strategy. It is true, of course, that the White House and the DOE have been careful to emphasize their desire to rely on market forces. But their understanding of that term has so far been broad enough to include proposals to subsidize domestic oil and gas production, and solar research. In this post-Reagan era the willingness to let markets provide answers seems to be waning. Most politicians, after all, are rivals of markets: the more we rely on the latter, the less power we confer on the former.

Existing Energy Resources

The alternative to the development of a new grand strategy is to use existing tools to make markets work better. Prices should be made to reflect the economic cost of each energy source, and regulators should be required to allow investors a reasonable chance to earn profits commensurate with the risks that they take. Also, monopoly positions should be challenged, so that competitive suppliers of energy—oil, nuclear, coal, and conservation technologies—can have equal access to markets. Subsidies, be they for Senator Dole's corn farmers or the greens' solar advocates, should be eliminated.

"In this post-Reagan era the willingness to let markets provide answers seems to be waning."

This freeing up of markets is not a national energy policy, as that term is generally understood. It is a program to permit private entrepreneurs to go about the business of satisfying consumer needs, confining regulation to those areas in which effective competition is not feasible. (Just as economic deregulation of airlines coexists comfortably with stringent safety regulation, so reliance on market forces can coexist with safety and other rules for nuclear plants, coal mining, and the use of oil tankers.) A program to make markets work better is, in short, a means of making a centrally constructed energy plan, or strategy, unnecessary.

Such a market-improving series of steps is intrinsically superior to any government-instituted energy plan. This is so not because government officials are less able than their private-sector counterparts, but rather because the mistakes that the government makes are carved in stone: it takes time, sometimes years or even decades, before Congress or a government agency even recognizes that it has made a mistake, let alone remedies it. Private-sector errors, on the other hand, are less enduring. We do not have to wait for the businessman to discover his own errors, because his competitors will discover his errors for him. If an oil-company executive sends inadequate supplies to a particular region, that will merely create an opportunity for his competitors; they will try to capitalize on it by increasing their supplies to the very region that the first executive undersupplied. If private-sector investors back a technology that does not work, they go bust; if the government backs one, more taxpayer money can always be conscripted to continue the program.

Recent history supports this view. Even the imperfectly competitive and partly regulated markets that constitute our energy economy have provided effective cures for many of the problems that began with OPEC's unsheathing of "the oil weapon" in the early 1970s. The grip of OPEC has been loosened because the extortionate prices set by the cartel induced consumers to use less oil, and producers to develop alternative supply sources. The oversupply in the electric-power business is being whittled away by demand growth and by asset write-downs, the latter resulting in demand-stimulating lower prices. The natural-gas glut is being worked off, ever so slowly, and contractual problems hanging over from the days of price regulation are being resolved. The environmental costs of fossil-fuel consumption (and, some say, more) will be reflected in prices when the new Clean Air Act comes into effect. And plans are slowly being drawn up by utilities and independents to build plants to meet future electricity needs. As a sailor, President Bush should know that, in such circumstances, steady as she goes is a much wiser course than a lurch to port.

The Bush Administration's National Energy Strategy Will Work

James Watkins

About the Author: *James Watkins is secretary of the U.S. Department of Energy (DOE). The DOE played a major role in developing the proposed National Energy Strategy unveiled by President George Bush in February 1991.*

In calling for his secretary to read him history, Frederick the Great would command, "Bring me my liar." Such aversion to seeing the truth in history is characteristic of many of the more vocal groups concerned with energy policy today. These groups seem to have little interest in reflecting on relevant history and learning from past experiences. But if we are to succeed, we must learn from the past and forge realistic policy approaches that will work in our democratic society.

With the release of the president's National Energy Strategy, the debate over energy policy has begun in earnest. Already the voices of the past are coming back to haunt us, proclaiming that there is only one way to affect American energy consumption: government edict. The voices call for taxes, mandates, regulations, and controls, all designed to make Americans do what government thinks is good. Yet history argues otherwise. The price control and allocation schemes of the seventies and the supply shortages they induced are not among the shining moments of energy policy history.

President Bush directed the Department of Energy to take a different approach. Rather than intervene in the free market, we've devised a strategy to harness the market's strength. Rather than choose between energy goals and environmental objectives, we have sought to balance them. Rather than more regulation, we are seeking more competition in every energy sector, thereby increasing fuel choices and reducing consumer costs. Rather than demand that Americans change their life-styles and make do with less, we are advancing the technologies that will lead us into the kind of energy future we want: more environmentally benign, with the economic growth necessary to maintain and enhance the American standard of living.

In developing this new energy policy, we reached out to the American public. Over some eighteen months we heard from 400 witnesses in eighteen separate hearings. We studied some 22,000 pages of testimony in a comprehensive effort to understand how Americans think our future energy needs should be met. Some called for conservation; some called for greater energy efficiency; others called for alternative fuels, advanced transportation technologies, and greater use of renewable fuel sources. Each group was and continues to be sure that its answer is the one "silver bullet" that will meet all our energy challenges.

"Rather than choose between energy goals and environmental objectives, we have sought to balance them."

They are wrong. None of these measures can *single-handedly* solve our complex energy problems. We will need *every one* of these initiatives, and more, to meet our future energy needs.

Past decisions have steadily eroded our energy choices. Our current energy system has been

James Watkins, "Fueling the Future," *Omni*, May 1991. Reprinted by permission of *Omni*, © 1991, Omni Publications International, Ltd.

shaped by decades of laws and regulatory mandates without regard for their cumulative impact. Project by project, decision by decision, we have eliminated one energy option after another and seriously compromised our overall energy and economic security.

Just consider: Frontier areas for oil and gas exploration in this country are largely closed to developers. Government at all levels has made decisions on nuclear power that make it virtually impossible for this technology to serve the nation. Multiple demands on water use have led us to sacrifice our hydropower potential. And to assuage local opposition, we have pushed the construction of needed new refineries overseas.

"Our goal is to maintain a healthy . . . free market so that all fuels and technologies can contribute to a growing U.S. economy."

By default, we have made imported oil the energy source of choice. It seems that no risk, however minor, is acceptable. Yet ironically, by refusing to accept minimal risks associated with expanding and diversifying energy supplies, we create greater risks by increasing our vulnerability to foreign oil producers.

Command and Control

Many in the old school of energy policy would like to institute command and control measures to reduce oil use and oil imports dramatically. There is no question that we must wean ourselves away from oil. But given oil's overwhelming importance to our economy, we must move wisely. In order to move away from oil without causing serious economic dislocation, we must introduce market incentives to stimulate development and use of alternative fuels and advanced energy technologies. Equally important, we must remove regulatory barriers that constrain existing alternatives like hydropower and natural gas. Our goal is to maintain a healthy, responsive free market so that all fuels and technologies can contribute to a growing U.S. economy.

Our nation faces serious energy challenges in the years ahead. We do not intend to compromise our future by repeating past mistakes. We will heed the lessons of history and build upon the foundation laid by the National Energy Strategy for a cleaner, more secure energy future.

Chapter 4: Preface

What Alternative Energy Sources Should Be Pursued?

Increased public awareness of the problems of fossil-fuel use has resulted in increased interest in alternative forms of energy. When discussing these energy alternatives, the term *renewable energy* is frequently used. Renewable energy sources are virtually unlimited in their availability. These include the sun, winds, ocean tides, rivers, and heat from beneath the earth's surface.

For centuries, humans observed the power of these natural phenomena and invented ways to harness their power. For example, the ancient Romans, Chinese, Egyptians, Phoenicians, and Greeks all used solar energy to dry crops and to evaporate saltwater to make salt. The Egyptians and many others harnessed the power of the winds to sail their ships, and in later times, the Persians and Chinese used windmills to pump water for field irrigation.

These early uses of renewable energy, however, were inefficient and unpredictable. Solar power was effective only at certain times of the day and in certain seasons when the sun's rays were strongest. Wind power was unreliable because of changes in the velocity and direction of the wind. And water power could be used only by those who had access to rushing streams.

The discovery of fossil fuels such as gas, coal, and oil in the nineteenth century made energy much more efficient and versatile. With the invention of the steam engine and internal combustion engine, the use of fossil fuels quickly overtook all other fuel sources and directly led to the Industrial Revolution in the late 1800s. The Industrial Revolution brought prosperity and progress, but it also brought problems. Cities became blackened with coal dust, and the air and water became polluted with toxic particles. A decrease in the use of coal eventually reduced its polluting effects. But the subsequent increased use of oil for internal combustion engines added pollutants to the air. In addition to these environmental problems, people also began to realize that fossil fuels were finite. This was especially evident when two major oil shortages in the 1970s interrupted America's fuel supply.

Some experts propose a return to the energy sources of the past as a way to avoid the polluting effects of fossil fuels. Scientists and engineers are applying modern technology to make these sources more efficient. Today, solar rays can be collected by photovoltaic cells and stored in batteries for an extended period of time. This method helps California's solar energy plants produce approximately 1 percent of the state's electricity. California is also the primary site for wind turbines. These now use improved blades to harness the wind more efficiently than traditional windmills. Thousands of slender wind turbines at three "wind farms" in California generate enough electricity to supply three hundred thousand homes.

Renewable energy sources still pose some problems, however. Despite new technology, wind power is limited because it is restricted to regions with steady winds. Some solar energy plants require large areas of open land for their hundreds of ground-based solar panels and reflectors. Another drawback to renewable energy sources is that they cannot compete with low-cost fossil fuels. In 1988, for example, a southern California electric utility purchased wind-generated electricity for almost eight cents per kilowatt-hour. In comparison, the utility produced electricity from oil, coal, and nuclear power for one-fourth that cost. Similar problems plague the solar energy industry. On average, photovoltaic electricity costs five times as much as conventional electricity. As long as the cost of these alternative energy sources remains high, consumers will continue to prefer inexpensive fossil fuels.

The search for new energy sources that are efficient, clean, and cost-effective continues. Only the future will reveal whether the U.S. can successfully harness the power of renewable energy sources to meet human needs.

Renewable Energy Sources Should Be Developed

Christopher Flavin and Nicholas Lenssen

About the Authors: *Christopher Flavin is vice president and senior researcher with the Worldwatch Institute, an independent, nonprofit research organization that focuses on global environmental problems. Nicholas Lenssen is a research associate with the institute.*

The end of the fossil fuel age is now in sight. As the world lurches from one energy crisis to another, fossil fuel dependence threatens at every turn to derail the global economy or disrupt its environmental support systems. If we are to ensure a healthy and prosperous world for future generations, only a few decades remain to redirect the energy economy. . . .

To stabilize the climate, the world soon will have to reduce its consumption of fossil fuels; this entails not only improving energy efficiency but also developing major new energy sources. Problem-plagued nuclear technologies clearly are not ready to play this role. Indeed, nuclear expansion has now come to a halt in many nations.

The alternative is as obvious as the sunrise: energy from the sun and other renewable resources. The technologies are at hand to greatly expand the use of renewable energy in the next few decades. . . .

Renewable energy resources are actually far more abundant than fossil fuels. The U.S. Department of Energy estimates that the annual influx of accessible renewable resources in the United States, for example, is more than 200 times its use of energy, and more than 10 times its recoverable reserves of fossil and nuclear fuels. Harnessing these resources will inevitably take time, but according to a new study by U.S. government scientific laboratories, renewables could supply the equivalent of 50 to 70 percent of current U.S. energy use by the year 2030.

Contrary to popular belief, renewables—primarily biomass and hydropower—already supply about 20 percent of the world's energy. Biomass alone meets 35 percent of developing countries' total energy needs, though often not in a manner that is renewable or sustainable in the long term. And in certain industrial countries, renewables play a central role: Norway, for example, relies on hydropower and wood for more than 50 percent of its energy.

"Renewable energy resources are actually far more abundant than fossil fuels."

Steady advances have been made since the mid-seventies in a broad array of new energy technologies that will be needed if the world is to greatly increase its reliance on renewable resources. Indeed, many of the machines and processes that could provide energy in a solar economy are now almost economically competitive with fossil fuels. Further cost reductions are expected in the next decade as these technologies continue to improve. As leading solar scientists, Carl Weinberg and Robert Williams, wrote in *Scientific American*: "Electricity from wind, solar-thermal and biomass technologies is likely to be cost-competitive in the 1990s; electricity from photovoltaics and liquid fuels from biomass should be so by the turn of the century." The pace of deployment, however, will be determined by energy prices and government policies. After a period of neglect in the eighties, many governments are now supporting new energy technolo-

Christopher Flavin and Nicholas Lenssen, "Beyond the Petroleum Age: Designing a Solar Economy," *Worldwatch Paper 100*, December 1990. Reprinted with permission of the Worldwatch Institute.

gies more effectively, which may signal the beginning of a renewable energy boom in the years ahead.

Direct conversion of solar energy will likely be the cornerstone of a sustainable world energy system. Not only is sunshine available in great quantity, it is more widely distributed than any other energy source. Solar energy is especially well suited to supplying heat at or below the boiling point of water (used largely for cooking and heating), which accounts for 30 to 50 percent of energy use in industrial countries and even more in the developing world. A few decades from now, societies may use the sun to heat most of their water, and new buildings may take advantage of natural heating and cooling to cut energy use by more than 80 percent.

Solar rays are free and can be harnessed with minor modifications in building construction, design, or orientation. In Cyprus, Israel, and Jordan, solar panels already heat between 25 and 65 percent of the water in homes. More than 1 million active solar heating systems, and 250,000 passive solar homes, which rely on natural flows of warm and cool air, have been built in the United States. Advanced solar collectors can produce water so hot—200 degrees Celsius—that it can meet the steam needs of many industries. Indeed, using electricity or directly burning fossil fuels to heat water and buildings may become rare during the next few decades.

"Solar energy will likely be the cornerstone of a sustainable world energy system."

Solar collectors, along with other renewable technologies, can also turn the sun's rays into electricity. In one design, large mirrored troughs are used to reflect the sun's rays onto an oil-filled tube that produces steam for an electricity-generating turbine. A southern Californian company, Luz International, generates 354 megawatts of power with these collectors and has contracts to install an additional 320 megawatts. The newest version of this "solar thermal" system turns 22 percent of the incoming sunlight into electricity. Spread over 750 hectares, the collectors produce enough power for about 170,000 homes for as little as 8¢ per kilowatt-hour, already competitive with generating costs in some regions.

Future Solar Technologies

Future solar thermal technologies are expected to produce electricity even more cheaply. Parabolic dishes follow the sun and focus sunlight onto a single point where a small engine that converts heat to electricity can be mounted, or the energy transferred to a central turbine. Since parabolic dishes are built in moderately-sized, standardized units, they allow for generating capacity to be added incrementally as needed. By the middle of the next century, vast areas of arid and semiarid countryside could be used to produce electricity for export to power-short regions.

Photovoltaic or solar cells, which convert sunlight into electricity directly, almost certainly will be ubiquitous by 2030. These small, modular units are already used to power pocket calculators and to provide electricity in remote areas. Within a generation, solar cells could be installed widely on building rooftops, along transportation rights-of-way, and at central generating facilities. A Japanese company, Sanyo Electric, has incorporated them into roofing shingles.

Over the past two decades, the cost of photovoltaic electricity has fallen from $30 a kilowatt-hour to just 30¢. The forces behind the decline are steady improvement in cell efficiency and manufacturing, as well as a demand that has more than doubled every five years. These cost reductions mean that in rural areas, pumping water with photovoltaics is already often cheaper than using diesel generators. Solar cells are also the least expensive source of electricity for much of the rural Third World; more than 6,000 villages in India now rely on them, and Indonesia and Sri Lanka also have initiated ambitious programs.

Photovoltaics, because of their lower projected cost, might eventually take over the central generating role of solar thermal power. By the end of this decade, when solar cell electricity is expected to cost 10¢ a kilowatt-hour, some countries may be turning to photovoltaics to provide power for well-established grids. By 2030, photovoltaics could provide a large share of the world's electricity—for as little as 4¢ a kilowatt-hour.

Another form of solar energy, wind power, captures the energy that results from the sun's unequal heating of the earth's atmosphere. Electricity is generated by propeller-driven mechanical turbines perched on towers located in windy regions. The cost of this source of electricity has fallen from more than 30¢ a kilowatt-hour in the early eighties to a current average of just 8¢. By the end of the nineties, the cost is expected to be around 5¢. Most of the price reductions have come from experience gained in California, which accounts for nearly 80 percent of the world's wind-produced electricity. Denmark, the world's second-largest wind energy producer, received about 2 percent of its power from wind turbines in 1990.

Wind Power

Wind power has a huge potential. It could provide many countries with one-fifth or more of their electricity. Some of the most promising areas are in northern Europe, northern Africa, southern South America, the U.S. western plains, and the trade wind belt around the tropics. A single windy ridge in Minnesota, 160 kilometers long and 1.6 kilometers wide, could be used to generate three times as much wind power as California gets today. Even more productive sites have been mapped out in Montana and Idaho.

Living green plants provide another means of capturing solar energy. Through photosynthesis, they convert sunlight into biomass that, burned in the form of wood, charcoal, agricultural wastes, or animal dung, is the primary source of energy for nearly half the world—about 2.5 billion people in developing countries. Sub-Saharan Africa derives some 75 percent of its energy from biomass, most of it using primitive technologies and at considerable cost to the environment.

"A single windy ridge in Minnesota . . . could be used to generate three times as much wind power as California gets today."

Many uses of bioenergy will undoubtedly increase in the decades ahead, though not as much as some enthusiasts assume. Developing nations will need to find more sophisticated and efficient means of using biomass to meet their rapidly increasing fuel needs. With many forests and croplands already overstressed, and with food needs competing for agricultural resources, it is unrealistic to think that ethanol distilled from corn can supply more than a tiny fraction of the world's liquid fuels. And shortages of irrigating water may complicate matters, especially in a rapidly warming world.

In the future, ethanol probably will be produced from agricultural and wood wastes rather than precious grain. By employing an enzymatic process, rather than inefficient fermentation, scientists have reduced the cost of wood ethanol from $4 a gallon to $1.35 over the past 10 years, and expect it to reach about 60¢ a gallon by the end of the nineties. Within a few decades, however, liquid fuel from biomass will be at a premium as oil production declines.

More efficient conversion of agricultural and forestry wastes to energy could boost biomass energy's role in the future, particularly in developing countries already reliant on this source. Wood stoves that double or treble today's efficiency levels already exist, and better designs are under development. For modular electricity generation, highly efficient gas turbines fueled by biomass can be built even at a very small scale. Some 50,000 megawatts of generating capacity, 75 percent of Africa's current total, could come from burning sugarcane residues alone. In the future, integrated farming systems, known as agroforestry, could produce fuel, food, and building materials.

Hydropower now supplies nearly a fifth of the world's electricity. Although there is still ample growth potential, particularly in developing countries, environmental constraints will greatly limit such development. Small-scale projects are generally more promising than the massive ones favored by governments and international lending agencies. Smaller dams and reservoirs cause less social and ecological disruption. In deciding which hydropower resources to develop, issues such as land flooding, siltation, and human displacement will play an important role. These considerations will likely keep most nations from exploiting all of their potential.

"Hydropower now supplies nearly a fifth of the world's electricity."

Another important element of a renewable-based energy system is geothermal energy—the heat of the earth's core. This is not strictly a renewable resource, however, and it needs to be carefully tapped so as not to deplete the local heat source. Since geothermal plants can produce power more than 90 percent of the time, they can provide electricity when there is no sun or wind.

Geothermal resources are localized, though found in many regions. Worldwide, more than 5,600 megawatts' worth of geothermal power plants have been built. El Salvador gets 40 percent of its electricity from the earth's natural heat, Nicaragua 28 percent, and Kenya 11 percent. Most Pacific Rim countries, as well as those along East Africa's Great Rift Valley and around the Mediterranean, could tap geothermal energy. Virtually the entire country of Japan, for example, lies over an enormous heat source that one day could meet much of the country's energy needs.

While fossil fuels have been in storage for millions of years, renewable energy is in constant flux—replenished as the sun shines. While not a constraint in the near future, the intermittent nature of sunshine and wind means that the large-scale use of renewables will need to be backed by some form of energy storage. Indeed, biomass energy and hydropower are the only forms that can be stored easily. Developing new and improved storage systems is therefore one of the key challenges in building a sustainable energy economy.

Heat below the boiling point of water can be stored in simple devices that rely on water, bedrock, oil, or salt. Thermal storage systems pump heat captured on sunny summer days through these substances, and then extract the heat when it is needed, such as on a cold winter night. Such systems can recover as much as 85 percent of the heat originally captured. Already, some 30 large solar-storage installations have been built in Europe, including 10 district heating systems in Sweden. District heating traditionally employs a central fossil-fuel-burning plant that delivers steam or hot water to neighboring buildings; the Swedish plants, however, use stored sunlight to supply heat to nearby schools, office buildings, and apartments.

Renewable Energy Storage

Storing electricity is a greater challenge. Pumped hydroelectric storage systems—which elevate water to a reservoir, then drop it through a turbine to produce electricity—are now used in some regions. Though their use is growing, pumped storage systems will likely be limited by the availability of sites and by environmental objections to dam building. Another less disruptive alternative is a storage system that uses electricity to compress air into an underground reservoir. When power is needed, the air is released, heated up, and forced through a turbine. As with pumped-hydro, compressed-air storage systems can achieve about 70 percent efficiency. A

290-megawatt system is already operating in Germany. Another technology, superconducting magnets, could offer highly efficient and inexpensive electricity storage, according to scientists, but it will not be ready for at least several decades.

"Hydrogen can also be used to generate electricity without producing nitrogen oxides."

Battery storage is a more flexible alternative. Home photovoltaic panels can be hooked up to batteries, as can utility-scale wind or solar plants. Batteries could also play a role in transportation, without greatly increasing electricity demand. If electric cars were used for one-quarter of U.S. auto travel, total electricity use would rise only 7 percent. At today's electricity prices, electric cars are already competitive with gasoline-driven ones in terms of fuel price. The challenge is to reduce the cost and extend the range of batteries beyond the current limit of 125 kilometers. During the early nineties, several major auto companies are scheduled to introduce electric vehicles.

Several new batteries are being developed. One cell that has been tested, the sodium sulfur battery, is more efficient, more compact, longer lasting, and lighter than current lead-acid models. But it requires further improvements before commercial use, including cheaper ways to keep the battery hot enough to function properly.

Hydrogen is the strongest candidate for large-scale storage. It is the cleanest burning fuel, producing only water vapor and small amounts of nitrogen oxides. These emissions can be reduced with lower combustion temperatures and nearly eliminated with specially designed catalytic converters. Hydrogen also can be burned in place of petroleum, coal, or natural gas.

The chemical industry currently produces hydrogen from fossil fuels, but it can also be made by electrolysis: splitting water molecules into hydrogen and oxygen with an electric current. German and Saudi engineers are developing electrolysis systems powered by electricity from photovoltaic cells. Proponents of solar hydrogen envision huge desert photovoltaic farms connected by pipelines to cities. Hydrogen can be stored in metal hydrides—metal powders that naturally absorb gaseous hydrogen, and release it when heated—or in pressurized tanks or underground reservoirs, thus providing a readily accessible form of energy.

Hydrogen can also be used to generate electricity without producing nitrogen oxides by chemically combining it with oxygen in a fuel cell. Hydrogen fuel cells are 70 percent efficient and could be used in hydrogen-powered electric cars. Internal combustion engines, by comparison, rarely convert even 25 percent of gasoline to usable energy, while standard power plants operate at about 35 percent efficiency. . . .

Using Energy Efficiently

Vastly improved energy efficiency—along with being intrinsically important in any effort to move away from fossil fuels—is the key to making a sustainable energy system work. If a home's electricity needs are cut by two-thirds, for example, the investment cost for a rooftop photovoltaic power plant could be halved. Similarly, a highly efficient electric car would go further and would need smaller batteries than a less efficient one, reducing its cost and weight. Thus, the development of more energy-efficient technologies is as crucial to the viability of an economy based on renewable energy as the solar technologies themselves.

Geothermal Energy Is an Abundant and Clean Energy Source

Augusta Goldin

About the Author: *Augusta Goldin is an author of science books for children and has written syndicated science features.* Small Energy Sources, *from which this viewpoint is excerpted, is her sixteenth book.*

Geothermal energy for our use! Say the words and people envision the huge steam-driven power plants in such countries as Italy, New Zealand, and the United States. But these developments, spectacular and profitable though they may be, are not likely to be duplicated to any great extent. There are just too few large-scale geothermal deposits that are hot enough and pressurized enough to blast roaring steam from wells and boreholes. However, the planet's geothermal potential in its *small-scale* sources is so great that it could provide much of the energy we need for millennia.

Although these small sources are hidden below the surface of the Earth, their heat energy breaks through at times. It explodes from volcanic craters and steam vents. It bubbles in molten lava. It glows in hot dry rocks. It simmers in high-temperature aquifers and occasionally leaks out in boiling springs. It even lurks in the soil and rocks under your feet.

Consider the heat energy in geothermal water that can be used as it comes out of the ground. Long before Columbus discovered America, Maoris were cooking their food in New Zealand's geothermal springs, Romans were swimming in geothermally heated pools in England, and ingenious Icelanders were piping hot water to their wintry huts. In the Middle Ages, embattled Tuscans were warring over the commercial possibilities of the hot mineral-laden springs in northern Italy, and since then such French towns as Claudes Aygues, Dax, and Aix-les-Thermes have been distributing near-boiling water to their inhabitants for domestic use. In the industrialized United States, some fine private buildings have also been geothermally heated since the nineteenth century. One of the first, in Boise, Idaho, was Moore House, built in 1883 and later hooked into the town's heating system, which included four hundred homes. Nor should we overlook the nearby 1902 Mott Mansion, which, large though it is, has never required much supplemental heat. Even in the coldest month of 1979, when temperatures hovered around zero, the fuel bill for that twenty-eight-room house was forty dollars.

Although small geothermal deposits are widely scattered, you're not likely to find any unless you happen to live along the boundaries of the Earth's crustal plates or the Ring of Fire that girdles the Pacific. It's in these places that most of the world's untapped hot reservoirs are located.

"There is so much geothermal energy, so many ways to harness it."

Where does all this heat come from? It comes from the molten magma deep in the Earth's interior. The top of this magma rises at times, heats the solid rock layer above it, and occasionally breaks through. Some of this heat is conducted to the next layer, which consists of porous rock through which water trickles and drips. Naturally, this moving water also heats up, and the result? A hot-water aquifer! If this aquifer is not too far below the surface and if its temperature seems to be satisfactory, it has the potential for small-scale development for the production of low-cost electricity—twentieth-century style.

To pinpoint such hot-water sources, prospectors scout around in vans laden with instruments and cameras. Geologists and hydrologists follow, to identify possible sites for development. If their reports are encouraging, the technical crew moves in, sinks exploratory boreholes, and monitors the temperature range, chemical composition, and flow rate. Then, if the findings are promising, engineers and technicians tap the source with drilled wells, then position the pumps, and install a small on-site power plant.

Hot-Water Systems

Hot-water sources that are within 300 to 600 feet (92 to 184 meters) below the surface can be found on all the continents. There brine simmers at temperatures ranging from less than the boiling point of water—212° F (100° C)—to something like 600° F (315° C). Under such conditions, developmental possibilities beckon and may be turned into realities with the use of appropriate technology.

Current appropriate technology includes hot-water systems that generate electricity without using boilers and without raising steam.

Here are two of those systems: the *flash* cycle, which operates with high-temperature fluids that are greater than 300° F (150° C); and the *binary* cycle, which operates with moderate temperature fluids—190° to 300° F (90° to 150° C).

It's the high-temperature flash cycle that is at the heart of the flash steam system. Here the technical crew pipes the hot geothermal fluid to a flash tank, whereupon, the pressure being reduced, it flashes—vaporizes to steam. The part that doesn't vaporize (called the condensate) is either reinjected into the ground or drawn off to heat swimming pools or greenhouses and the like. The flashed steam rushes off to drive the turbine generators. The spent steam, on leaving the turbine, is chilled by the condenser, turns to water, is pumped to the cooling tower, and is finally disposed of into the injection well. This is a continuous process that keeps on producing electricity as long as the pipe keeps delivering the self-flowing brine to the flash tank.

And it's the moderate-temperature binary cycle that is at the heart of the organic rankine cycle system. Here the geothermal fluid is used in conjunction with a secondary working fluid; hence the term *binary*. This secondary fluid is, typically, a liquid refrigerant, a fluorocarbon or halocarbon (such as the commercial Freon) that moves through the cycle in a closed loop and never comes into contact with the brine. . . .

Such hot-water technologies were virtually unknown prior to the middle of the twentieth century. The first flash-cycle plants began operating, with rated capacities ranging from three to fifty-five megawatts, in New Zealand in 1958, the USSR and Japan in 1967, Iceland in 1968, Mexico in 1973, El Salvador in 1975, the Philippines in 1977, the Azores in 1979, and the United States in 1980. These and other countries have since developed a considerable number of successfully operating flash-cycle plants. . . .

"Reykjavik, once blackened by the smoke pouring from its chimneys, is now the cleanest city in the world."

Binary-operated plants with their slightly more involved technology lagged somewhat in development. Although small operations had been churning out low-cost electricity in China, it wasn't until the eighties that installations began appearing in the western United States as industrialists began learning what the engineers at the Electric Power Research Institute (EPRI) already knew—that with moderate-temperature fluids, the binary system is the most efficient and reliable way to generate power at the lowest cost for the utilities. The binary system, they say, requires only two-thirds of the geothermal fluid that a flash system needs to produce the same amount of electricity. In addition, they say, with the binary, all the brine is returned to the source, so the supply is not depleted. Note, however, that because the temperature of the fluid drops during the operation, that in itself is a kind of depletion that needs to be considered. So convincing has EPRI been that the USDOE [United States Department of Energy] and the San Diego Gas and Electric Company are investing heavily in binary-cycle plants in California's Imperial Valley.

Some field engineers have become so intrigued by the fact that geothermal hot water can be used to generate electricity that they have invented portable, packaged power plants and are installing them in settled areas far from utility grids, where the cost of flying in diesel fuel is prohibitive.

"Earth heat is an abundant and inexhaustible source of energy."

Gary Shulman, president of the Geothermal Power Company (New York), is one of those inventors. In a personal communication, he described his flash-cycle system, which he calls the Monoblok. "This," he wrote, "is a portable turbine that is skid-mounted and can be installed quickly on a drilling pad to convert flashed steam to electricity. It contains all the necessary systems and controls, similar to an automobile that is ready to drive when purchased." The first Monoblok—installed in Dieng, Java, in 1980—taps a reservoir with an estimated potential of one thousand to two thousand megawatts. This small power package, left unattended and unwatched except for occasional maintenance, has been generating two megawatts of electricity in an area where the people had never before seen an electric light.

Besides "Monobloking" geothermal wells in Indonesia, Shulman plans to tap the Soufrière Volcano on the Caribbean island of St. Lucia with five- and ten-megawatt installations that he expects will provide electricity at three cents a kilowatt hour. "Compare this," he says, "with what people there are now paying for conventionally produced electricity—fifteen to twenty cents a kilowatt hour! Can't you just see rural electrification and small industrial developments in their future?"

Mitsubishi of Japan and Ormat Ltd. of Israel are also responding to the challenge of hot-water development by installing modular, mobile power package units in communities that are sitting on geothermal deposits in Mexico, the Philippines, and the western United States. Particularly interesting as examples of modular expansion are the installations in Sulphurdale, Utah, and Steamboat Springs, Colorado.

There is so much geothermal energy, so many ways to harness it, and at the same time such a global need for low-cost power! In a paper presented at the 1985 International Symposium on Geothermal Energy, Dr. Ronald DiPippo of Southeastern Massachusetts University stated that some 188 power plants—propelled by dry steam, wet steam, and/or hot water—were already working away in seventeen countries and generating almost five thousand megawatts of electricity. Currently, some scientists are predicting that worldwide production will top 10,000 megawatts by 1995. . . .

Direct-use Systems

Clearly, the greatest geothermal success story to date has been the production of electricity from large steam-dominated reservoirs, but the story now in the making is the exploitation of liquid-dominated sources with flash- and binary-cycle technologies. The third, longer-range success story, however, may well turn out to be the *direct utilization* of low- and moderate-temperature water, interest in which has recently been revitalized because the cost of fossil-fuel energy is so high. In addition, new sensing devices now enable scientists to locate these resources with a fair amount of speed, new technologies help engineers drill those relatively shallow wells with dispatch, and inexpensive, off-the-shelf components are widely available.

But caution is needed. No matter how plentiful the source, how desirable the flow rate, and how steady the geothermal temperature, the fact remains that the chemistry of the fluid needs to be taken into consideration. Some brines are so highly mineralized and corrosive that they ruin the pumps and pipes. Some engineers have handled the problem in such a way that the hot brine never leaves the well. Only the heat is pumped out. . . .

"Expert scientists say [geothermal] sources could provide 75 percent of our energy needs in the foreseeable future."

If, however, the geothermal fluids are just barely mineralized, they may safely be pumped from the well right through the heating and hot-water systems of residential properties, commercial establishments, and district projects.

In the residential category, Oregon would appear to be the showcase for direct-use systems. If, for example, you lived in a private house in Klamath Falls, you'd probably have your very own geothermal well simmering in your front lawn, complete with two U-tube down-hole heat exchangers suspended in it. There are more than 550 such residential properties in that town, with inexpensive systems installed and long paid for. The owners, consequently, are enjoying cost-free space conditioning and hot water—amenities that will be theirs for perhaps twenty or thirty years, after which the well temperature will very likely have dropped and the heating systems will be in need of a supplemental boost.

In the commercial category, the Oregon Institute of Technology campus has been a resounding success since 1964. The geothermal fluid there, being only slightly mineralized, is pumped from three wells to and through the systems in eleven college buildings. The annual costs for cooling and heating those buildings and for providing them with domestic hot water runs to some $35,000. This includes maintenance, salaries, equipment replacement, and repair, plus the cost of electric power for pumping. It all computes to 6.3 cents a square foot (.0929 per square meter). Using conventional fuel, the estimated annual costs would run to about 60 cents a square foot.

In the matter of district heating and hot water, Reykjavik in Iceland is the world's most successful example of direct utilization. The geothermal fluids, drawn from hot springs and drilled wells in Warm Springs Valley, serves the homes, as well as commercial and institutional buildings, greenhouses, and outdoor swimming pools the year round. And the cost? About 70 percent of what it would be were the inhabitants still firing their boilers with fossil fuel. As for the environment, Reykjavik, once blackened by the smoke pouring from its chimneys, is now the cleanest city in the world. . . .

In considering geothermal energy sources, one that is rarely mentioned is *earth heat.* This is solar energy that has been absorbed by and stored in the Earth's crust throughout millennia. It is stable, it is consistent, and it hovers below the frost line at a temperature of some 45° to 56° F (7° to 13° C). In northern latitudes, the frost line generally settles some 4 to 5 feet (about a meter and a half) below the surface. Accordingly, farmers dig their root cellars below the frost line so they can store apples and potatoes, carrots and beets safely in the winter.

Earth Heat

Earth heat is an abundant and inexhaustible source of energy, but you are probably thinking that at 45° to 56° F, it is too low to bother with in twentieth-century fashion. Not so. Earth heat is providing low-cost space conditioning, by way of *heat pumps,* to some twenty thousand buildings in the United States and tens of thousands more in Canada, Sweden, France, and other countries. Even on the coldest days, when the ground is blanketed with five or six feet of snow and ice, such a system can extract low-temperature heat from the earth and deliver it at a higher temperature to wherever it is wanted. Countless private homes, barns, schools, hospitals, condominiums, and industrial structures are being space-conditioned twelve months a year without burning a single stick of wood, lump of coal, drop of oil, or cubic foot of gas or drawing on geothermal steam or hot water. Instead they use earth heat.

"Countless . . . structures are being space-conditioned twelve months a year."

Driven by electricity, heat-pump systems can extract thermal energy from the air, the water, or the Earth itself. It's when these systems go underground that earth heat is transferred most consistently, reliably, and economically through closed-loop coils—pipes that are made of metal or plastic. These are buried in one of two configurations: as horizontal earth coils or as vertical ones. In either case, the coils are filled with a working fluid, which is water or a solution of water and a form of antifreeze such as propylene glycol, depending on the depth of the frost line.

In the horizontal system, the coils are buried in shallow trenches some 3 to 7 feet (roughly 1 to 2 meters) deep. Considerable land space is needed, about twice the size of the building that's to be heated. If land is at a premium, then the loops are installed in drilled vertical shafts that may extend downward 60 to 700 feet (about 18 to 215 meters).

Whether the mode is horizontal or vertical, heat pumps that are coupled with closed-loop earth coils circulate the heat that was absorbed by the working fluid and may also augment its temperature. The warmed water in the earth-coil system flows alongside the refrigerant tube in the evaporator (the heat exchanger). The refrigerant, which is a liquid fluorocarbon, absorbs heat from the water in the coil. Now, since the fluorocarbon has a low boiling point, it quickly begins to vaporize, flows to the compressor, and gets squeezed, whereupon its temperature rises. Then that high-temperature, high-pressure gas is pumped to the condenser, where it gives up its heat to the building as a fan blows across the condenser coils. As the building warms up, the gas cools down, changes back to its original state—a liquid refrigerant—and is moved along to the expansion valve, where the temperature and the pressure are further reduced. Then we're right back where we started—with a low-temperature, low-pressure refrigerant that is pumped back to the evaporator, and the process starts all over again.

A building can be cooled just as easily with the same closed-loop earth-coupled system if it is equipped with a reversing valve. All you need to do is flip the switch. . . .

Future Prospects

In general, the future of heat pumps for earth-heat extraction depends on two factors: climate and the cost of electricity. In areas where the climate is not too harsh, the future looks more than promising because, despite high installation charges, new and improved technology has so reduced trenching and drilling expenses, the cost of components, and electricity usage that payback time is only three to five years. After that, except for minimal repair, the system practically calls out, "Come—help yourself to free earth heat."

Accessible earth heat, geothermal steam, and low-temperature hot water constitute formidable sources of energy. Expert scientists say these global sources could provide 75 percent of our energy needs in the foreseeable future, and there are now enough success stories to prove it can be done.

Solar Energy Is a Promising Energy Alternative

Steven Ashley

About the Author: *Steven Ashley is an editor for* Mechanical Engineering *magazine. Ashley is a former energy editor for* Popular Science, *a monthly science magazine.*

"Topping the thirty-percent-efficiency mark in solar cells is like breaking the four-minute mile in track," says Dan Arvizu, former supervisor of photovoltaics at the Department of Energy's Sandia National Laboratories in Albuquerque, N.M. Arvizu and I stand in a darkened Sandia lab room watching James Gee, a staff investigator, painstakingly position two thumbnail-sized rectangles on a copper pedestal. An actor in a spotlight, Gee works under a yellow shaft of noontime sunshine. It comes from a xenon lamp.

Gee fixes a pair of electrical probes to the small stacked squares, sits up, and disappears into the surrounding gloom. "I've just assembled our tandem solar cell," he says quietly. "Under concentrated sunlight it reaches a record thirty-one-percent efficiency level. The device is a sandwich: a gallium-arsenide cell made by Varian Associates in Palo Alto, Calif., and a crystalline-silicon cell fabricated at Stanford University.

"The gallium-arsenide cell on top absorbs the visible part of the solar spectrum, and the silicon cell on the bottom uses the transmitted infrared light that's left," Gee explains. "The idea behind tandem or stacked cells is that two or more PV [photovoltaic] materials can absorb the spectrum more efficiently than can a single material. This raises the efficiency of the whole," Gee says.

Steven Ashley, "New Life for Solar?" *Popular Science,* May 1989.

Though the record-setting cell is currently too costly to market, industrial researchers are using the same stacked-cell strategy to boost the efficiency of less-expensive commercial-grade solar cells, experts say. In the last few years, in fact, a range of solar-cell technologies has hit new performance marks as engineers apply insights derived from better understanding of photovoltaic materials and of semiconductor-device design.

Years of research in government, industrial, and university labs—much of which began in the days of the energy crisis—are bearing fruit, experts say. Besides demonstrating startling performance improvements, engineers are developing innovative manufacturing methods that may reduce production costs. Meanwhile, industry has begun operating new automated factory lines for producing lower-cost current-technology PV devices in larger volumes. And in the field, pilot-size solar power plants built years ago are showing that PVs can work for extended periods.

"A range of solar-cell technologies has hit new performance marks."

The dream of clean, abundant low-cost solar power for the future has been around for a long time: no fuel to burn, few parts to fix, little supervision required. It always was one of the brightest options on the energy horizon. . . . Now it's time to take the pulse of this key technology once more. I visited research labs around the country to ask experts to evaluate the state of the industry and to point out the experimental solar cells to watch.

"The basic problem with the solar industry is simple: High-performance solar cells cost too much," explains Eldon Boes, the photovoltaic projects manager at Sandia Labs, as we stroll into the facility's solar-array test field.

"We test PV modules here," he says. "With today's fuel costs you'd like to get module costs down to at least one to two dollars per peak watt—the power output in midday sunlight. Now it's more like four to six dollars."

The sun had cleared the mottled-brown mass of Sandia Mountain, and the lab's photovoltaic arrays—giant high-tech sunflowers on metal stalks—had raised their variegated foliage to the sun. Boes and I pass different varieties of solar collectors standing in rows around the fenced-in field like exotic crops in a garden plot. Several box- and barrel-shaped arrays sheathed in thin plastic lenses refract spectacular multifaceted rainbows from the glow.

Other dark-faced solar panels, armored in playing-card-size squares of indigo, blue, and black, lie in ebony rows like huge fallen dominoes. We watch as they shift silently toward the sunlight like sluggish seesaws. Volatile fluids within their hollow support frames vaporize gradually in the sun's heat as shadows shorten, overbalancing the panels by degrees as the sun passes. "Crystalline-silicon PV systems like those work quite well. That is, they convert a good portion of the light to electricity—maybe twelve percent—over a long period of time," says Boes. "The problem," he explains, "is that they're made out of many silicon wafers individually sliced from crystalline ingots, treated, and then soldered together into a module. This labor-intensive production process leaves the module too costly—maybe four to six dollars per peak watt—to compete with cheap fossil fuels."

Cheaper Solar Power

"So how do you make solar power cheaper?" Boes asks. "One highly effective way is to increase the solar-cell efficiency, as Gee and his colleagues and many other research groups are attempting. If you double the conversion efficiency, you more than halve the system cost." Besides requiring fewer modules, there are fewer support structures, and less wiring, land, and installation labor needed.

Every so often motors whir and click as they drive rows of ray-catching fresnel lenses in unison to follow the sun. With each sudden studied movement new spectral color splashes around the test field. Boes gestures toward the various lensed units and continues, "Another school of thought says: If good cells are so expensive, rather than covering your collectors with costly cells, cover them with cheap acrylic lenses to gather the light, then focus it on your cells and so increase the power you get from each cell." But concentrators, he adds, require motorized two-axis sun-tracking systems. This extra cost plus the need for direct sunlight makes them promising mainly for large-scale systems in areas such as the Southwest.

"Amorphous silicon can reach higher efficiencies by playing the stacking game."

"Another way," says Boes, "is to make inherently inexpensive PV materials that work pretty well and lower the cost that way. This avenue has successfully provided electricity for consumer electronics and small power systems." Boes points off to one side where four rows of black rectangular panels scored with delicate silver grid lines lean back catching some rays. "Sovonics/ECD, Chronar, Solarex, and Arco Solar make these thin-film amorphous-silicon modules. The manufacturing costs are potentially low—on the order of a dollar per watt—because they need little photovoltaic material and they can be mass-produced. But their conversion efficiencies are also rather low—around five or six percent—so you need a lot of them to produce a reasonable amount of power, and that ups the cost." The Department of Energy [DOE], through the Solar Energy Research Institute in Golden, Colo., he notes, funds research on many thin-film PV technologies.

It's not overly surprising to learn that many industry observers pick thin-film amorphous (non-crystalline) silicon technology as having the inside track to utility-scale power generation. Since its introduction in 1979, amorphous silicon has become the fastest growing market segment and now accounts for two-fifths of the global trade. "It's a natural progression," says David Carlson, general manager of Solarex's thin-film division in Newtown, Pa. "Amorphous silicon has been successful because you can lay down thin coatings onto inexpensive substrates relatively easily and cheaply. What is holding it back is its low efficiency."

The chemical-vapor-deposition apparatus used to make such cells varied little among the factories and labs I visited. Using large vacuum chambers filled with precisely controlled mixtures of silicon and hydrogen gases and small quantities of various doping agents such as boron, technicians deposit micron-thick films on glass or stainless-steel substrates. Radio-frequency coils inside the reinforced tanks generate electromagnetic fields that ionize the gases. The ionized atoms in the resulting hot plasma plate out one by one on the substrate surface, which is kept at several hundred degrees Celsius. Layers with the special opto-electronic properties are produced by adding different dopants to the gas mixture during progressive deposition cycles. Other deposition stations along the production line lay down additional anti-reflective coatings as well as films that provide electrical conduction and protective encapsulation.

"Performance can be improved by increasing the amount of light entering the cell."

Major amorphous-silicon manufacturers are cutting labor costs and raising fabrication rates by installing large production lines that use automated equipment such as robot materials handlers, conveyor belts, scribing lasers, and computerized precision process controls. For instance, Chronar Corp. in Princeton, N.J., is testing a new line designed to produce 10 *megawatts* of amorphous PVs annually.

The enhanced capability of this line was demonstrated when it fabricated a monolithic PV module 2.5 by 5 feet—among the world's largest.

Like the record-setting DOE-Varian-Stanford cell, amorphous silicon can reach higher efficiencies by playing the stacking game. "Stacking amorphous silicon with certain silicon-germanium alloys is one of the widely investigated ways to raise efficiencies," Solarex's Carlson explains. "The silicon absorbs the visible light while the silicon germanium collects the infrared light, which increases the total light the cell collects.

"We calculate that the practical efficiency limit for amorphous silicon alone is maybe fifteen percent," he says. "The potential efficiency of a stacked amorphous and silicon-germanium cell is twenty-five percent or more. That leaves a lot more room for improvement. And that's why people are studying this approach. The real beauty of it is that all you have to do to make these stacked cells is to add a few more deposition chambers to a standard production line," Carlson adds.

Achieving Efficiency

"The key now," he concludes, "is to develop improved silicon-germanium alloys with the electrical properties that will allow a solar cell to achieve those high-efficiency values. This is where a lot of the proprietary research is going on now. It's not far off; we expect to see high-efficiency stacked amorphous silicon cells in the early '90s."

Many experts point to Energy Conversion Devices in Troy, Mich., as the probable leader in advanced amorphous-silicon technology. ECD has been manufacturing six-percent-efficient tandem panels for two years.

"We solved the fundamental problems of making stacked cells long ago," says Subhendu Guha, vice president for photovoltaics. "Right now we deposit six layers of amorphous silicon onto one-thousand-foot rolls of fourteen-inch-wide stainless-steel sheet in an automated process. One three-layer cell absorbs mostly blue light. As we deposit the other silicon cell, we play around with the process conditions so that it absorbs more yellow and red light. The tandem layers are more efficient."

But in ECD's view the real payoff comes when a film of silicon-germanium alloy is added to the layer cake. Notes Guha: "This layer absorbs the red and the infrared light, which further extends the cell's absorption range. We've demonstrated cells with 13.7-percent efficiency and foot-square modules rated at 8.4 percent—the world's highest for amorphous silicon. We're convinced that we can make ten percent triple-stack amorphous modules by the end of 1989," he notes. "ECD engineers are designing the machines to produce them right now."

Light Degradation

Amorphous silicon might not have had to go the stacked-cell route had a major barrier to higher efficiencies been overcome. Says Guha: "Amorphous silicon's big problem is that its efficiency degrades by about ten to fifty percent when it's exposed to light. This is known as the Staebler-Wronski effect, named for those who reported it first in 1977." In fact, it's one of the most investigated phenomena in PV research.

"Amorphous silicon has no long-range order. That is, the silicon atoms are jumbled throughout the material in no particular order," Guha explains. "The absence of crystalline structure means that many silicon atoms do not bond strongly with their nearest neighbors as they would in a crystal." This occurs because many atoms are missing from the positions they'd inhabit in a regular crystal lattice.

"According to the conventional wisdom," he continues, "those weak bonds break when they are exposed to sunlight, forming defects called dangling bonds that can serve as sites where electron-hole pairs recombine." When this occurs, he says, the pairs never make it to the electrodes, thus cutting current output. Annealing of the silicon by the sun's heat removes new dangling bonds, Guha adds, and the opposing processes—defect formation versus annealing—eventually reach an equilibrium state. The result is a moderate decline in the original efficiency, but it's enough to hinder amorphous silicon's progress.

"Copper indium diselenide has created a good deal of excitement among PV researchers."

Though the Staebler-Wronski effect can be minimized by avoiding defects by adding the right amount of hydrogen to "tie up" the dangling bonds, and by making the films in the stacked cells so thin that the electron-hole pairs are never far from the electrodes, no one has so far managed to eliminate it completely.

The rapid progress of a new polycrystalline photovoltaic material called copper indium diselenide has created a good deal of excitement among PV researchers. CIS, as it's known, was just another PV candidate a few years ago, but engineers at Arco Solar in Camarillo, Calif., have demonstrated a square-foot-sized panel with an 11.1 percent efficiency, the highest of any large-area thin-film device.

"The rapid evolution of CIS is truly remarkable," says Larry Kazmerski, principal scientist at the Solar Energy Research Institute (SERI) and an early investigator of the material. CIS offers many intrinsic benefits: "It has a wide band-gap window, meaning it absorbs much of the solar spectrum," explains Kazmerski. "In addition, it has extremely high optical-absorption properties, which means a thin layer can absorb a lot of light, thus lowering material costs. And unlike amorphous silicon, its photovoltaic properties are stable."

Charles Gay, president of Arco Solar, is also enthusiastic about CIS: "Copper indium diselenide is more forgiving of impurities as well as variations in composition and in thickness than other thin films," he tells me during a visit to the company's offices in California's Ventura County. This characteristic, he says, makes it easier to scale up the manufacturing process Arco researchers are using now.

Gay raises a glass-enclosed plastic-framed PV module from the table. "This is a CIS panel," he says. The material behind the glass plate looks black in the light. "We had a small-area cell with 12.5-percent efficiency in 1985. Several years of hammering away at a variety of problems resulted in this module. It has a micron-thick layer of CIS coated with a 0.03-micron layer of cadmium sulfide. We have these being tested at SERI now."

Then he picks up a semi-transparent orange-red module and places it on top of the CIS module. "Our plan is to stack this thin-film silicon module mechanically on a CIS module to get even higher efficiencies," Gay says. "The top silicon layer collects the bluish light and the CIS on the bottom uses the red. So far we've gotten an efficiency of 15.6 percent from a square-inch-size tandem cell."

International Solar Electric Technology, an Inglewood, Calif.-based firm founded by former Arco Solar scientist Vijay K. Kapur, has plans to make CIS cells with a low-cost electroplating technology.

"Engineers began optimizing solar-cell designs by applying strategies borrowed from the semiconductor industry."

As work continues on CIS, yet another thin-film polycrystalline PV material, cadmium telluride, is receiving attention, says SERI's Kazmerski. Although its light-absorption characteristics are well-matched to the solar spectrum and it can be manufactured inexpensively, there have been problems with cadmium telluride. It can be difficult to make electrical connections with the material, and its electrical properties are often hard to control. Also, some observers worry about the disposal of the toxic cadmium after the cells stop functioning.

Nevertheless several companies are making cadmium-telluride cells. Japan's Matsushita Battery Industrial Co. uses a low-cost printing method to lay down thin layers of cadmium telluride and cadmium sulfide on a ceramic base. The plant cranks out one million cells a month. Britain's BP Solar is also said to have invested heavily in cadmium-telluride technology.

Semiconductor Strategy

In the United States both Ametek Inc. in Paoli, Pa., and Photon Energy in El Paso, Texas, are exploring low-cost mass-production methods for cadmium telluride. Photon Energy has built a low-cost pilot production line that has fabricated square-foot panels of cadmium telluride that are 7.3 percent efficient. These panels are now under test at SERI. Ametek engineers have produced an 11-percent-efficient cell by electrodeposition, says Peter Meyers, Ametek's director of applied materials. A one-foot-square cadmium-telluride panel made by a proprietary method will be brought to market soon, he says.

High-performance high-cost single-crystal silicon has been the workhorse of the industry since its inception, says SERI's Kazmerski. But in the early 1980s no one expected substantial efficiency gains beyond the then-current 11 or 12 percent. At that point, however, higher-purity crystalline silicon became available, and engineers began optimizing solar-cell designs by applying strategies borrowed from the semiconductor industry. The outcome was that single-crystal silicon cells with efficiencies of up to 22.6 percent have been developed.

Researchers such as Stanford University's Richard Swanson and Martin Green of the University of New South Wales in Australia are leaders in this effort, Kazmerski says. One semiconductor-design strategy, for example, says that performance can be improved by increasing the amount of light entering the cell. Swanson's and Green's designs minimize light blockage by shrinking front-surface electrical contacts and by coating the top with improved anti-reflective materials. They also have front and back coatings of tin oxide that "passivate," or tie up dangling silicon bonds formed by the end margins of the cells.

Swanson's Stanford group has developed "point-contact" PV devices for lens-equipped concentrator arrays that have demonstrated 28.2-percent efficiency. These devices have no light-obscuring front contacts at all. Positive and negative point contacts line the reflective back plate. Manufacturing costs for these complex cells remain an area of research, however.

Allen Barnett, another leader of the effort to apply optimization strategies from the semiconductor industry to solar cells, is general manager of the Astropower division of Astrosystems Inc. in Newark, Del., and professor of electrical engineering at the University of Delaware. He is working on making cells from thin layers of lower-quality polycrystalline silicon. This approach, he feels, will reduce manufacturing costs.

"It's true that PV technology seems to be in a good state right now."

"Starting out in 1980, I decided that the way to get the required cost and performance from PVs was to come up with a method to make thin-film polycrystalline silicon in a continuous fashion," Barnett recalls. "My fellow researchers told me, 'Yes, Allen, it's the right question, but it is too hard to solve.' "

Barnett says that the optimization equations with which he works told him that good results could be obtained if he were to follow several rules: Use thin films so that generated electron-hole pairs are never far from the contacts and so they do not have to travel far before they are collected by the circuit contacts. Use light-trapping techniques such as textured back surfaces: "The photons bounce sideways off the back so that they think they are going through a thick piece of silicon. These elongated pathways enhance the light's chances of getting absorbed." Also, find a substrate that provides both mechanical support and back-surface conduction while also matching the thermal-expansion properties of silicon. And finally, form "benign" grain boundaries between the crystals that don't halt the transport of electrons and holes.

The result: a 15.7-percent-efficient thin-film polycrystalline silicon cell.

Today Barnett reports that he's on his way to commercializing a process in which 100-micron-thick layers of polycrystalline silicon are grown on special low-cost ceramic substrates 10 centimeters on a side. His proprietary process involves first depositing a reflective metallurgical barrier on the ceramic panels, and then, on top, a thin solution of silicon and tin. A conveyor belt brings the package through a furnace that allows planar polycrystalline silicon to grow on the ceramic. "In my process," Barnett says, "a line of ten-by-ten-centimeter ceramic panels comes through like a train of boxcars."

He hopes to have a manufacturing line producing at a rate of one million wafers per year. These one-watt devices will have at least 8.5 percent efficiency, he says. Unlike thin-film cells, however, the finished polycrystalline cells must still be screenprinted with electrodes and soldered together like other crystalline silicon cells. Nevertheless, Barnett claims significant cost savings for his process. His goal is a 14-percent-efficient cell that costs about $2 per peak watt.

A Growing Industry

Whatever the advances in solar photovoltaic technology prove to be, the PV industry remains a big question mark. Consider this assessment by Don Schueler, manager of Sandia's solar-energy department and long-time observer of the field: "It's true that PV technology seems to be in a good state right now. Technical advances seem to be rolling in at a faster rate, and industry is adopting them as they come along.

"But the health of the industry is problematical," he warns. "Though the producers are selling their products in niche markets, few companies make money, and some long-term financial backers such as the oil companies want out of the business. Meanwhile, both governmental and industrial funding for research and development is declining. . . . Investment in solar research is expensive. That's hard to justify when a company is concerned about its bottom line. I'm afraid that this is a short-sighted view."

"The Solar Age may dawn sooner than we know."

Nevertheless, there's reason for hope, because new factors have entered the solar equation. For instance, following the Chernobyl nuclear-power-plant disaster, European countries such as West Germany and Italy started to sink more money into solar research. West German government spending rose to $47 million. Similarly, the Japanese government is continuing its heavy support for solar power in hopes of developing energy self-sufficiency and export profits. In 1988 it spent $54 million on PV research.

Further, a sea change in attitudes about utility power generation may be in the offing. "Some of the more progressive energy planners," Sandia's Schueler says, "are starting to recognize the 'total' cost of energy production. That is, what is the entire cost to society of choosing one power-generation technology over another?" For example, the electric bill we pay today for nuclear-power-generated electricity does not include the cost of storing spent radioactive fuel or of decommissioning old plants. "Planners estimate that the true cost is about double what we pay."

Similarly, should the nascent fears of global warming prove valid, the total societal cost of fossil-fuel energy production would be staggering. If this belief grows, perhaps consumers would be willing to pay more for solar power, leading utilities to take the plunge.

But don't count on it. In the real world market forces determine power generation choices, and for solar power, that comes down to demonstrated high-performance and long-term reliability at a low cost.

If the solar photovoltaics industry has not yet become the multi-billion-dollar business some observers had predicted several years ago, there's no question that it's growing. Close to $400 million worth of PV products—over 30 megawatts—were sold in 1988. The products are varied: rural power-generating systems for Third World villages and other sites far from utility grids; remote electrical power supplies for water pumps, electric fences, and communications and navigation equipment; stand-alone power systems for consumer products like calculators, radios, appliances, walkway lights, and auto-battery chargers.

The Solar Age

As for large-scale PV power generation, it remains pretty much where it was—in the future. Solar-cell manufacturers say that they are relying on the current "intermediate-market" products to keep the industry afloat until falling PV prices and rising energy costs meet at the "crossover point"—where solar cells become competitive with other sources of energy. Then, they say, the big utility sales will materialize (initially, for "peaking" power plants to generate extra power during periods of high demand) and the costly investment in high-volume production lines will be justified. After that, high-output grid-linked PV panels for the roofs of businesses and residences will become more affordable.

Predictions of the crossover point vary widely. Some say it's close, others say it's years away. Much depends on the newest generation of PV technology. If performance keeps improving and prices continue to fall, however, the Solar Age may dawn sooner than we know.

Wind Power Is a Clean and Renewable Energy Source

Shawna Vogel

About the Author: *Shawna Vogel is an associate editor for* Discover, *a monthly science magazine.*

When the Organization of Petroleum Exporting Countries [OPEC] tripled the price of oil in the mid-1970s, many of the world's petroleum-importing countries took a closer look at such renewable sources of energy as the sun, the wind, and the tides. In the case of wind energy, the spurt of research activity led to some rapid improvements in a centuries-old technology. But the interest was short-lived. With the price of oil now down . . . private companies are discouraged from investing in windmills; and with the federal budget deficit dominating the political debate, the U.S. government, for one, has scaled back the wind-power research program that was born of the energy crisis.

Nevertheless, wind power is far from dead. It's alive and well in California, where it produces one percent of the state's electricity. In Hawaii and Quebec, wind turbines of record size came on-line in 1988 and are performing well. And while governments on this side of the Atlantic are bowing out of the wind-power business, the United Kingdom is making a major new commitment. Its Central Electricity Generating Board has announced plans to build three wind farms, covering 750 to 1,000 acres apiece, in southwestern England, northern England, and western Wales. The farms will each be equipped with 25 hundred-foot-high turbines putting out 300 to 500 kilowatts apiece; all told, they'll generate enough electricity for about 15,000 people.

Shawna Vogel, "Wind Power," *Discover*, May 1989, © 1989 Discover Publications. Reprinted with permission.

Although wind power is generally thought of as environmentally benign, the British electricity board is braced for complaints from the public—not only because the wind farms may interfere with radio and television signals but also because they may be seen as ugly blots on the green and pleasant land. "Our windiest sites are also our most cherished sites," notes Jill Sandland of the electricity board, and selecting sites for farms is a problem.

At the same time, the board plans to build the world's first offshore wind turbine. The experimental turbine, which will have an output of 750 kilowatts, will be mounted on a steel tripod sunk into the floor of the North Sea three miles off the coast of Wells in East Anglia. There it will be less obtrusive, and it will also benefit from the faster winds that prevail offshore. On the other hand, building and operating a turbine at sea (an underwater cable will channel the electricity back to land) will be more expensive than building one on land.

"[Wind power] is alive and well in California, where it produces one percent of the state's electricity."

Both the offshore turbine and the wind farms are to be completed in the early 1990s, and at that point the British will decide whether the technologies will be developed further. If all goes well, says Lord Marshall, chairman of the electricity board, "we might hope that early next century the wind will give us a thousand megawatts of economic electricity generation"—about the capacity of a large nuclear power plant.

Hawaii is also a windy place, and unlike Britain it doesn't have its own oil or coal. Nearly

90 percent of its electricity is generated from imported oil. That explains why Hawaiian Electric Industries, a holding company with interests in energy as well as real estate development, has long been interested in wind power. In 1988 the company acquired a showpiece: the world's largest horizontal-axis wind turbine. Now supplying power to 1,200 homes on Oahu, the 3.2-megawatt propellerlike turbine was built by Boeing and NASA [National Aeronautics and Space Administration]. The rotor, which measures 320 feet from tip to tip, is mounted on a 200-foot-high tower.

"In Hawaii and Quebec, wind turbines of record size came on-line in 1988 and are performing well."

The device is not only big but also technically advanced. On most wind turbines the rotor is allowed to spin at only a single fixed speed, which is dictated by the requirements of the electric generator it is driving. When the wind gusts, the generator prevents the rotor from spinning faster. The result is a lot of torque on the drive shaft and a potentially damaging spike in the power output.

In contrast, the speed of the rotor on the Hawaiian turbine is allowed to change gradually as the wind speed changes. The turbine can operate more efficiently at high speeds because the rotor absorbs some of the energy of a gust, thereby smoothing out spikes in the torque and the power. It also operates at lower wind speeds (as low as 12 miles an hour) than other large turbines do. In general the smoother operation reduces the stress on the machinery. "That means you can reduce the size of the components," says Dan Suehiro of Hawaiian Electric. "The drive shaft doesn't have to be as thick, and the gears don't have to be as beefy. You reduce the amount of steel you have to haul up the tower."

A turbine that started full-time operation in 1988 at Cap Chat, Quebec, avoids both these problems. It has a vertical rather than a horizontal axis—it looks like an eggbeater rather than a propeller—and it too is the largest of its kind. Its 308-foot central column is flanked by two bowed blades that are 205 feet apart at their farthest extension. The blades can be moved by wind from any direction. On the other hand, because they are mounted near the ground, they are generally exposed to slower winds than the blades of a large horizontal-axis turbine.

The 3.6-megawatt Cap Chat turbine, owned by the engineering firm Lavalin, is the last to be constructed under the Canadian National Research Council's alternative energy program. The program was started in 1975, after OPEC boosted oil prices, and canceled in 1985. The Hawaiian turbine too is the end of a line: it is the last of six turbines built under a U.S. program whose goal was to demonstrate the feasibility of large wind turbines. That goal has now been met.

Wind Power

Yet in the absence of tax incentives for wind-power developers—the U.S. government eliminated those incentives in 1985—the immediate economic outlook for wind power is poor, largely because fossil fuels are so cheap. Wind is also less reliable than coal, oil, or natural gas. And what it saves in environmental pollution can't easily be translated into dollars.

In the long run, though, the price of fossil fuels will almost certainly go up, and Edgar DeMeo of the Electric Power Research Institute in Palo Alto, California, is hopeful about the long-term future of wind power. "If the wind-turbine developers can survive the next few years," he says, "my feeling is the market will open up. In three or four years we'll have a five-cent or perhaps a four-cent per kilowatt-hour machine. In areas where there are good winds, it will be reliable and compatible with the utilities' needs. The situation today won't continue indefinitely. The only question is when it's going to stop."

Harnessing the Power of the Oceans Can Provide Energy

Marlise Simons

About the Author: *Marlise Simons is a writer for the daily newspaper,* The New York Times.

On a rocky headland bashed by the Atlantic Ocean, engineers are finishing a wave power installation that they hope will prove an important stepping stone toward much larger projects in the rush for renewable and clean energy.

With a simple device in a wave-washed gully, they are tapping some of the immense energy stored in the swell of the ocean and converting it into electricity. They use waves to produce an airstream that drives a turbine. In the coming weeks, the turbine will be linked to a generator that will supply all the current needed by a small community.

This small station is the simplest effective wave-power device created so far, its engineers say. Its team of engineers from Queens University in Belfast is part of the growing group of experts who argue that there are no inherent obstacles to harnessing the enormous energy potential of the sea.

The pilot plant at Islay, an island on the west coast of Scotland, is expected to produce a modest average of 40 kilowatts a day, enough for the nearby little village, which now uses diesel fuel. But specialists say units of this size can already be multiplied for island communities or desalinization plants.

Along the stormy west coast of Europe, several recent studies note, the ocean swell is so mighty and near constant that it could theoretically provide places like Ireland, Scotland and Portugal with more electricity than they need. In deeper offshore waters where waves can have five times as much power, some experts say, the yield could be far greater.

Although small-scale experiments with wave energy have been going on for several decades, more ambitious projects for commercial quantities of energy have long lingered on the drawing board. Advocates argue that even when projects were technically and economically feasible, progress has been unnecessarily slow, because they have lacked political and financial support.

But interest in wave power has warmed as the politics and pollution of oil and nuclear energy have become more troublesome. The Persian Gulf crisis, experts say, is now likely to increase support for all forms of renewable energy.

The European community commission has for the first time included wave power among alternative energy sources earmarked for research funds. Norway, Britain, India and Japan have also committed new funds.

"Man has always been afraid of the power of the sea but I think we are now in the early stages of tapping it," said Trevor Whittaker, a naval architect who heads the wave power team at Belfast University.

"The ocean swell . . . could theoretically provide places like Ireland, Scotland and Portugal with more electricity than they need."

The new wave plant sits on a bleak promontory on Islay, a place that feels the full blast of the Atlantic breakers on most days of the year. In the gully, the team set down a hollow concrete chamber, its front wall partly open to the sea. As a new wave arrives, usually about every 10 seconds, the rising water in the chamber pushes the air into a vent holding a turbine, which drives a generator.

The receding wave and the dropping water level create a vacuum so that air is sucked back in and continues to drive the turbine. This is possible because Alan Wells, an engineer in Belfast, designed a rotor whose blades are lifted like the wings of an airplane by rising air so that it can spin independently of the direction of the air. American one-direction rotors works on a different principle, with blades pushed onward by air.

"The technology is so simple, it's amazing a wave energy device was not invented years ago," said Stephen McIlwayne, a team member. He has spent many hours on Islay, timing and gauging the swell of the ocean, which inspires him to speak of wave energy on an epic scale.

"You watch that tremendous power all the way down the coast, and it makes you think of the hundreds of megawatts that are being dashed on the shore and not being used," he said. "The ocean is like a big battery, a huge collector and we can tap it in many places. It's just a matter of time."

Other Nations' Wave Experiments

Mr. Whittaker cautioned against too much short-term optimism because of the fate of some earlier plants. A Norwegian tower, which worked on a similar but more complicated principle, was unhinged from the rocks and destroyed in a storm in 1988. India planned for the world's largest wave power station in Madras by setting a series of turbines into a harbor wall. The project was canceled when waves proved not to be strong enough year round.

Japan has experimented with devices powering buoys and lighthouses. It is now finishing a breakwater for a fishing village that holds devices to produce 40 kilowatts a day. A Portuguese group wants to start a pilot project on the Azores.

But in Britain a bitter dispute has erupted involving the most sophisticated device designed so far, the so-called Salter duck.

Although Britain had moved faster in wave energy research than other nations, the government cut off most funds in 1982, after an unfavorable report that presented wave energy as more costly than wind energy and instead recommended more work on wind.

Stephen Salter, a professor of mechanical engineering at Edinburgh University, has fought the decision ever since. Dr. Salter invented the duck, a machine he said could generate large volumes of energy offshore.

"Ocean waves are the most concentrated form of renewable energy we know, and it is continually created."

Dr. Salter charged that the negative report was prepared by experts from the British nuclear agency who strongly favored nuclear energy. He said they had a vested interest in distorting his data and cost estimates. The controversy grew as one consultant who worked on the report testified that his key conclusions were "changed and even reversed" against his will.

The experts responsible for the report have denied that there was any conflict of interest and insisted that the duck was not cost effective. Other engineers have argued that the fight arose because the two sides used widely different assumptions for calculating costs.

Wave Energy Studied

The controversy has become the subject of hearings in Parliament, and Britain's Department of Energy has now begun a new review of the viability of wave energy. Some oceanic engineers hope that this will lead to a new lease on life for Salter's duck.

Tested in a large wave tank at the University of Edinburgh, the duck, a canister, generates electricity as it moves with the waves, nodding up and down on a spine. Dr. Salter, a robotics specialist, acknowledged that his device would be costly and that it might take years of work before it could go offshore.

Although different countries are now researching a range of devices, Mr. Salter said there were still only two basic methods to make energy from waves: "Either you use something rigidly fixed, which must resist every force the wave throws at it, like a boxer who has his head against the wall. Or you have a more complex device that dodges out of the way like the boxer who moves on his feet. The second is far more costly, but it will work in deep water and yield far more energy."

Installations using underwater concrete baffles to focus waves toward a narrow point on shore are another example of the fixed-device class, but they have not had great success so far.

Many experts say they believe that wave power at best can provide only a small part of overall energy needs. Mr. Whittaker believes it is too early to make that judgement.

"Technically there is enough energy along our coasts to power Britain," he said, "but at this point to deliver 1,000 megawatts to the consumer, you need a wave power plant with machines spread out along 30 miles at some distance from each other, much like underwater fenceposts."

With currently available technology, wave power is useful only for small communities. Mr. Whittaker therefore sees the pilot plant on Islay only as a "way to improve the credibility of wave power and then move on to larger projects."

Like other advocates, he argues that the "technology is too young to judge it." He went on: "You could not have asked the Wright brothers to design a jumbo. If we had looked at the early motorcar or the plane this way, we would have axed the industry."

Britain, with its large stretches of coast facing the open Atlantic, is now making a survey of wave energy resources along its shores. Other countries, including Ireland, France, Spain and Portugal, are considering making similar studies. Such research includes looking for "hot spots" along headlands and other places where waves concentrate. "Waves bend and converge and focus much like light waves," said Mr. McIlwayne, "so the topography creates places of greater intensity."

"The ocean is like a big battery, a huge collector and we can tap it in many places."

Other new water-powered projects include setting ranks of barriers to take advantage of changing tidal water levels in the estuaries of coastal rivers. France already has one such tidal energy plant in the mouth of the Rence. But further plans in western Europe have been opposed by environmentalists who say the large barriers would be too disruptive for natural life.

Compared with solar and wind energy, Mr. Whittaker said, "ocean waves are the most concentrated form of renewable energy we know, and it is continually created." He continued: "Waves hitting us today could have been generated days ago somewhere else. The whole ocean acts as a collector of solar energy."

The Commission of the European Community will have more than $25 million available for research into alternative energy. "We think wave energy has a strong potential and deserves some of those funds," said Philippe Bourdeau, the commission's director for environmental and energy research.

Chapter 5: Preface

What Are the Alternatives to Gasoline?

Alternatives to gasoline have been available for many years. However, their use lagged until the 1970s, when oil shortages forced the U.S. to consider a reduction in oil imports. In particular, the use of methanol and ethanol gained much attention. The main issue concerning these fuels today is whether or not they are viable alternatives to gasoline.

Unlike gasoline, ethanol and methanol are not produced from refined crude oil. They are alcohol fuels. Methanol, or wood alcohol, is normally manufactured from either natural gas or coal. Ethanol can be derived from corn, sugar cane, and other plants. Proponents favor ethanol and methanol because the resources needed to produce them are abundant in the U.S. For this reason, advocates claim, investing in these fuels would enable the nation to reduce its dependence on foreign oil imports. As a result, the billions of dollars the nation pays to oil-rich countries would remain in the U.S. Says Daniel Sperling of the National Research Council, "From a purely economic perspective, we're insane not to pursue alternative fuels."

In addition to economic benefits, many people argue, developing alternative fuels would reduce the amount of air pollution produced by burning gasoline. Automobiles that use gasoline emit into the atmosphere harmful particles, such as hydrocarbons, nitrogen oxides, and carbon monoxide, and they contribute to smog. Methanol, advocates say, burns more cleanly than gasoline and cuts polluting emissions by half. In fact, many industries and federal and state governments, such as California, are converting their fleets of cars, trucks, and buses to methanol to comply with clean air standards.

However, many problems need to be solved for these fuels to compete with gasoline. For example, critics contend that ethanol and methanol decrease the performance of cars. A common complaint against ethanol is that it corrodes engine and fuel system parts. This would cause expensive automotive repairs for the consumer. Also, alcohol-based fuels are more expensive than gasoline, which dampens enthusiasm for their development. Opponents of methanol maintain that twice as much methanol is required to equal the mileage that gasoline offers. This means increased costs in automobile design because the size of fuel tanks would need to be doubled to offset the decrease in fuel efficiency. To make ethanol competitive with gasoline, state and federal governments have given gasohol producers more than $4.6 billion in tax exemptions since 1980. Without these exemptions, critics claim, ethanol would be far less developed and available than it is today.

In addition to being more costly and less efficient, alternative fuels are not as clean as the public is often led to believe, critics contend. In a California study, methanol was shown to emit slightly more hydrocarbons than gasoline. In other tests, methanol generated higher emissions of nitrogen oxide. Says critic Michael Fumento, a writer on alternative fuel topics, "All fuels cause pollution, and some 'clean fuels' are dirtier for some emissions than gasoline."

While the development and viability of alternative fuels is debated among energy policymakers, America's reliance on oil and gasoline continues. Many experts believe that lower costs and superior performance warrant the continued use of gasoline. Others contend that alternative fuels make more economic and environmental sense. The authors in this chapter discuss the alternatives to gasoline and their side effects.

Ethanol Fuel Can Replace Gasoline

Richard J. Durbin

About the Author: *Richard J. Durbin is a member of the U.S. House of Representatives' Transportation and Agriculture subcommittees. A Democrat, he has represented Illinois since 1982.*

Anyone who remembers the Arab oil embargo during the 1970s knows the price of dependence on foreign energy. The United States was importing 35 percent of its crude oil back then, and a temporary loss of Middle East petroleum led to long lines at gas pumps, inflation, and an economic recession.

That experience convinced Congress to help develop alternative sources of domestic energy. Ethanol is one of the success stories of that effort. It is a homegrown solution to not just the energy problem, but to several other troublesome issues as well.

Ethanol is a clean-burning alcohol fuel derived from grain (primarily corn in the United States), which typically is blended in a mix of 10 percent ethanol to 90 percent gasoline to produce "gasohol." It serves as an octane-booster much like lead, which has been sharply reduced in gasoline for environmental reasons. Ethanol blends are available in most states, particularly in the Midwest where most of the corn for ethanol is grown.

To encourage development of an ethanol industry, Congress passed a law in 1978 partially exempting alcohol fuels from the federal excise tax on gasoline. This legislation, which has been periodically extended, made the use of ethanol economically competitive, and the blend began appearing on the market in 1979. . . . The use of tax incentives to help create a new technology is not without precedent. Congress has used this strategy before with considerable success, most notably with the oil industry.

There are several major benefits from ethanol. Recent studies by the Congressional Research Service (CRS) and an independent research panel formed by the U.S. Department of Agriculture (USDA) showed that even a modest expansion of ethanol use in the United States would:

- Enhance national energy security by reducing our dependence on foreign oil.
- Help cut the federal deficit by reducing the cost of federal farm programs.
- Improve air quality by reducing tailpipe emissions of carbon monoxide from automobiles.
- Increase farm income and bring needed new jobs to struggling rural economies.

This combination of benefits led the President's Task Force on Regulatory Relief to encourage the use of "oxygenated" alcohol fuels (such as ethanol) as a fast and inexpensive way to help meet federal clean air requirements. In early 1988, the Environmental Protection Agency (EPA) announced new guidelines that allowed cities to meet federal clean air standards by using alternative fuels, rather than imposing severe driving restrictions on local residents.

"Even a modest expansion of ethanol use in the United States would . . . enhance national energy security."

That action follows 1987's decision by Colorado to mandate the use of oxygenated fuels during winter months to help reduce cold-weather air pollution problems in that state—primarily automotive carbon monoxide levels in the Denver metropolitan area. The Colorado Department of Health has estimated that the exclusive use of ethanol blends would reduce car-

Richard J. Durbin, "Ethanol: A Home-Grown Answer to a Host of Problems," *Forum for Research and Public Policy*, Winter 1988. Reprinted with permission.

bon monoxide emissions up to 34 percent; EPA projects a 25 percent reduction.

Because it is a relatively new and different product, ethanol has been blamed unfairly for a host of problems. Some major oil companies and dealers have suggested that alcohol blends might degrade engine performance or damage engine parts. An official of Chevron Oil Company, however, has testified before Congress on behalf of the American Petroleum Institute that his company has found no problems associated with the use of ethanol blends. Nevertheless, anti-alcohol advertising (by companies that have no stake in ethanol production and view ethanol blends as unwanted competition) have planted doubts in the minds of consumers.

The truth is that ethanol is a proven fuel: Domestic production has grown rapidly from 40 million gallons in 1980 to more than 800 million gallons in 1987. U.S. motorists have successfully driven 500 billion miles on ethanol-blended gasoline from 1980 to 1988.

Brazil's Fuel Policy

Brazil fuels its entire automotive industry on blended or pure ethanol produced from sugarcane. In Brazil, most gasoline stations have two pumps: One dispenses a 22 percent ethanol blend; the other dispenses pure ethanol. Nearly all new cars produced in Brazil are designed to run on pure ethanol. . . .

Despite the benefits of promoting an indigenous ethanol industry, U.S. ethanol producers face some potentially significant problems.

Most immediate is the drought of 1988, which destroyed an estimated 34 percent of the year's corn crop. With world oil prices still low, corn prices shooting up, and the corn surplus beginning to shrink, some small- and medium-sized ethanol producers will be forced to suspend production. The 1988 drought relief bill enacted by Congress generally limited federal aid just to farmers. If droughts persist, the ethanol industry could be hurt badly.

A longer-term problem is the scheduled expiration—on September 30, 1993—of the federal tax incentive to produce ethanol. Unless the current economic situation changes, the future of the U.S. domestic ethanol industry may depend on action by Congress to renew the tax exemption. . . .

"The exclusive use of ethanol blends would reduce carbon monoxide emissions up to 34 percent."

But despite the problems—both real and potential—ethanol is making a major contribution in this country. It could be making a much greater contribution, at a time when the United States is growing ever-more vulnerable to another energy crisis. We are currently importing about 40 percent of our oil, even more than in 1973. About 6 percent comes from one of the world's most politically unstable regions—the Persian Gulf.

According to one oil industry estimate, the true, unsubsidized cost of Persian Gulf oil is $140 a barrel because of the expensive U.S. armada we maintain in that region—far above the current spot price of under $20 per barrel. In that light, even the loudest critics should take another look at ethanol's benefits.

What would happen if the 1973 oil shortage was repeated? According to CRS projections, during a five-year period the United States would lose more than $700 billion in income, inflation would double, and unemployment would shoot up 25 percent. Past experience suggests this is not a far-fetched possibility.

Ethanol can help avoid that scenario, while helping farmers, the environment, and the economy. It is a homegrown solution to a host of problems and one that we can no longer afford to exclude from the nation's energy strategy.

Hydrogen Fuel Can Replace Gasoline

Jay Stein

About the Author: *Jay Stein is a designer of mechanical systems for custom homes and an author on environmental issues.*

As an alternative fuel for automobiles, hydrogen holds great promise. Frank Lynch, president of Hydrogen Consultants, Inc., of Littleton, Colorado, has been converting gasoline-powered engines to run on hydrogen since 1970. "First we take off the pollution devices and alter the fuel control system," he says. "All our engines are fuel-injected. We lose power to make them ultra-clean, so we put on a turbo-charger." An American Motors Gremlin converted by Lynch ran so cleanly that "there was less carbon monoxide and hydrocarbons coming out of the exhaust pipe than there was in the room air."

For nearly seventy years, Ben Jordan, a retired engineer living in Denver, has advocated the use of hydrogen fuels. In 1981, he set the land speed record for a one-liter GT coupe at the Bonneville Salt Flats' "flying mile." The following year, he spent $12.32 converting that same car to run on hydrogen. While making a dash across the salt flats in the transformed car, he was clocked at 132 miles per hour. To demonstrate the cleanliness of hydrogen fuel, Jordan likes to fill up a glass with condensed vapor from the tail pipe and take a swallow. He is currently converting his Model T Ford to hydrogen power.

On-board storage represents a major obstacle to the widespread use of hydrogen in motor vehicles. A Department of Energy (DOE) study concluded that "using liquid hydrogen requires more than twice the volume and four times the weight of an equivalent amount of gasoline." But with today's technology, the storage problem is not so overwhelming as to preclude the use of hydrogen-fueled vehicles. They can be found in locations where a premium is placed on keeping poisonous emissions to a minimum, such as congested European cities and underground mines. Lynch sees a bright future for hydrogen vehicles. "Let the technology take root where it makes sense now, and it will grow from there."

The mere mention of hydrogen fuel still arouses fears of explosions and uncontrollable infernos. Most experts in the field blame lingering memories of the 1937 Hindenburg disaster for the public's hydrogen phobia. These fears have some basis in fact, given that hydrogen has a much greater ignitability than gasoline and natural gas. Many people seem surprised to learn that hydrogen is routinely traded as a standard industrial commodity. Applications include welding, cooling electrical generators, making oils into shortening, and fueling rockets. Manufacturers in the United States produce 150 billion cubic feet a year with a safety record that Bob Gilardi of the Compressed Gas Association describes as good.

"As an alternative fuel for automobiles, hydrogen holds great promise."

Because of their inherent danger, equipping motor vehicles with hydrogen fuel presents a special challenge. Tankers of liquid hydrogen regularly cruise our roadways and railways with an excellent safety record. In at least one accident, a tractor trailer jackknifed on a wet road and spilled its hydrogen cargo, but the gas dissipated rapidly without igniting. By their nature, all fuels present some degree of danger. After nearly twenty years of experience, Lynch characterizes hydrogen as a "manageable safety risk

Jay Stein, "Hydrogen: Clean, Safe, Inexhaustible," *The Amicus Journal*, Spring 1990, © 1990, *The Amicus Journal*, a publication of the Natural Resources Defense Council. Reprinted with permission.

compared to conventional fuels."

As with any new technology, research is needed to overcome an array of technical challenges. DOE's solar hydrogen research program is funded at $2.5 million a year, according to Bill Hoagland, director. While he would like to see this increased, the program is lucky to be funded at all. In 1983, the program was shut down, and was only restarted in 1985.

Hoagland's boss, Henry Hubbard, director of the Solar Energy Research Institute, recently testified before the House Committee on Science, Space, and Technology. Describing the current research program he said, "I don't believe . . . it's adequately focused or really thought through." He later called the program "an orphan child" and suggested "a program should be conceptualized and defined." Hoagland has been charged to establish priorities and is emphasizing hydrogen production and distribution. "Until you have an economic source of hydrogen, nothing else matters," Hoagland says.

Three pieces of legislation under consideration in Congress call for increased funding for hydrogen research. Senator Tim Wirth (D-Colorado) has characterized the global warming trend as "the most important public policy issue of our lifetime." His National Energy Policy Act would allocate $590 million over a three-year period for hydrogen fuel and related research. Another global warming bill, introduced by Representative Claudine Schneider (R-Rhode Island), calls for spending $200 million over five years on renewable hydrogen research.

Both Wirth's and Schneider's bills draw heavily on legislation introduced by Senator Spark Matsunaga (D-Hawaii), the Senate's most ardent hydrogen advocate. Every year since 1985 he has introduced similar legislation. The bill for 1990 would allocate $55 million over five years for hydrogen production and use, and another $100 million for hydrogen-fueled aircraft research and development. All three bills have attracted significant support; Schneider's alone has 103 cosponsors.

Possible Drawbacks

But not everyone in Congress shares Matsunaga's enthusiasm. When Louis Ventre, Jr., counsel to the House Energy Research and Development Subcommittee, spoke at a recent workshop, he acknowledged the technology's benefits. He also noted "several significant drawbacks that tend to limit federal support." His concerns included problems with production and storage and the lack of an existing distribution system. "Clearly, the government could mandate socioeconomic changes to spur consumer demand," he said. "This seems unlikely, however, in the absence of a crisis which is the usual incentive for large government efforts."

"Tankers of liquid hydrogen regularly cruise our roadways and railways with an excellent safety record."

John O'M. Bockris of Texas A&M University, considered by many to be the guru of hydrogen fuel research, calculates that energy-related pollution costs the United States approximately $450 billion per year, or $1,800 a person. He estimates the cost of building a solar-hydrogen energy base over twenty-five years at $120 per person annually.

The question may not be whether we can afford the transition to a solar-hydrogen energy base, but whether we can afford not to make that switch. There is little doubt that the days of fossil fuels are numbered. Pollution-free hydrogen may be the dominant fuel in a future devoid of petroleum.

Methanol Fuel Should Replace Gasoline

Philip E. Ross

About the Author: *Philip E. Ross is a feature writer for the monthly science magazine* Popular Science.

When it's springtime in the South, cold weather doesn't come to you; you have to go to it. That's what Jim Hawkins did in March 1989 with a bunch of fellow engineering students.

"It was a typical college project, a month behind schedule, and we had to find a way to test our car for cold-starting," he said. "Teams from some of the other schools had cold rooms, but our only bet was to drive up Clingmans Dome, 6,600 feet up in the Great Smoky Mountains. It's the highest place in Tennessee that you can get to easily."

The 20 students from the University of Tennessee, in Knoxville, were working 50-hour weeks to convert a Chevrolet Corsica to run on a mixture of 85 percent methanol and 15 percent unleaded gasoline in a cross-country rally sponsored by General Motors and the Society of Automotive Engineers to promote alternative fuels.

With smog inundating many of our metropolitan centers, auto makers, oil companies, and federal regulators are looking at a whole range of automotive fuels that spew less pollution into the atmosphere. The best candidates for clean-air fuels are natural gas, methanol, ethanol, hydrogen, reformulated gasoline, and electricity. At this point, methanol appears the most likely one to make the trip soonest from the labs to the highways.

The Methanol Marathon wasn't just a school project. It was also a test of methanol as a practical automotive fuel under real-world conditions. A month after the student-prepared cars crossed the finish line, President Bush endorsed an air-pollution reduction plan that mandates the use of 1,000,000 alternate-fuel cars, the bulk of which will likely rely on methanol.

"Methanol's high octane lets it burn smoothly at higher compression than gasoline will tolerate."

Methanol is the top candidate due to its easy conversion from plentiful natural gas and reduced emissions of hydrocarbons and nitrogen oxides—both components of smog. How methanol-fueled cars will run; how much they'll pollute; where their fuel will come from; and how much it will cost are the questions on everybody's mind. The answers to these and other pressing questions will determine methanol's future as an automotive fuel. There's no better place to begin than at the starting line of the Methanol Marathon.

GM provided each college team with a Chevrolet Corsica equipped with a 2.8-liter V6 engine and a conversion kit that included a stainless-steel fuel tank and other parts specially designed to withstand the corrosive effects of methanol. Other sponsors provided student teams with access to dynamometers and diagnostic instruments to observe combustion and sniff out exhaust pollutants. The teams competed for points in average speed and fuel economy over the 1,100-mile rally route, beginning at GM's Technical Center in Warren, Mich., and winding through Toronto, N.Y., Delaware, and finishing at the University of Maryland's College Park campus. Other points were awarded for fastest acceleration and lowest noise and tailpipe emissions.

Hawkins says the University of Tennessee's victory, finishing ahead of 14 rivals, came from simple hard work rather than any technical tour de

Philip E. Ross, "Clean-Air Fuels for the 90s," *Popular Science*, January 1990.

force. Perhaps that is the greatest lesson the auto companies can learn from these neophyte engineers: "It was eighty percent attention to detail and doing reasonable, conservative things, like changing the gear ratio to conserve fuel in fifth gear," according to the student engineer.

"We knew a lot of schools would increase the compression ratio above the stock 8.9 to one," says Hawkins, who notes that methanol's high octane lets it burn smoothly at higher compression than gasoline will tolerate. "But we used turbocharging because it only works on full power. So on normal driving you get lower fuel consumption and more docile performance." These modifications sped the University of Tennessee car from a standing start to 500 feet in 8.1 seconds, just one-tenth of a second behind first-place Concordia College, which was also powered by a turbocharged engine.

"The Big Three auto manufacturers are already well along in developing the first generation of production methanol cars."

But the Tennessee team clinched its victory and a large portion of the $20,000 prize fund by taking first place in both emissions and fuel economy, at 19.9 mpg. "That doesn't sound very good," Hawkins allows, "but remember that's equivalent to more than thirty-eight miles per gallon of gasoline." You need nearly twice as much methanol as gasoline to go from here to there because it has a lower energy content: 65,000 Btu [British thermal unit] per gallon versus 116,000 Btu for gasoline. This is partially offset by methanol's better combustion efficiency, but not enough to avoid the need for a bigger fuel tank or a lot more pit stops. "You might as well get used to pumping twice the fuel into the car," says Hawkins.

These larger fuel tanks would likely take some space out of the trunk or the passenger compartment. And though engineers say some space up front could be saved by installing a smaller engine optimized for methanol's unique properties, there is no way to avoid sacrificing range, roominess, or both.

Likeliest Candidate

Huge fuel tanks are out of the question for the first generation of production methanol cars because their designs are based on existing gasoline-fueled models, such as the Chevrolet Lumina and Ford Taurus. Still, designers expect to be able to squeeze in a few more gallons by making tanks out of methanol-tolerant fluorinated plastics that can be molded to exploit every unused nook and cranny of the car's body.

But of greater concern is whether methanol, or any of the other alternate fuels for that matter, can make an impact on the nation's smog-shrouded cities. Says Karl Hellman, chief of emissions controls at the EPA's [Environmental Protection Agency] testing facility in Ann Arbor, Mich.: "Methanol, natural gas, and electricity are the leading candidates for clean-fuel cars—methanol is cited most as the example, but it doesn't mean other fuels can't do the job." Still, Hellman acknowledges that methanol seems the likeliest candidate for wide-scale application in the short run, though natural gas vehicles may make sense for high-mileage fleet users with central refueling stations.

The first methanol cars will be able to burn a variety of blends, from straight gasoline to M85—a 100-octane methanol blend containing 15 percent gasoline. That solves the problem of distributing a fuel for which there will be a small initial demand, but at the cost of optimizing performance for a particular blend. GM calls this solution "variable fuel," Ford calls it "flexifuel," and Chrysler calls it "gasoline-tolerant methanol," but they all work in a similar way. Key to adjusting for a variety of fuel mixtures is an electronic system that continuously tests the fuel. An optical or electrical-field sensor determines the gasoline-methanol mix, and feeds the data to the engine's control computer, which automatically adjusts fuel injection and spark timing to get the most out of each mixture.

Methanol, called wood alcohol because it was once made from wood chips, has an octane rating of 110 in pure form. That lets it burn without knocking at higher pressures than even premium gasoline can tolerate. "Racers have known this for years, beginning in boats—that's where I got my first experience with it," says Roberta J. Nichols, a senior engineer at Ford and an amateur racer on rivers and roads. "Its molecules are smaller than the hydrocarbons in gasoline, so you can squeeze more into the engine. A professor I knew at [the University of California at] Berkeley used to call it 'chemical supercharging.'"

She added that racers get more power by raising compression ratios as high as 12:1, exploiting pure methanol's high octane rating. The student-modified cars in the Methanol Marathon couldn't raise ratios so high, though, because they ran on M85.

There are reasons, unique to methanol's properties, why some gasoline needs to be added:

•Gasoline helps cold starts by raising the volatility of the fuel from the low level of pure methanol. Unlike methanol, gasoline is a complex blend of hundreds of compounds with different vaporization levels that can be adjusted to the latitude and season.

"From an environmental perspective, methanol does have some distinct advantages over gasoline."

•Gasoline keeps the fuel vapor in the tank space above the fuel line too rich to burn, while pure methanol can produce an inflammable environment. And special heat-absorbing grids need to be installed in the tank and fuel lines to interrupt the propagation of sparks because methanol conducts electricity all too well.

•Pure methanol burns with a dull blue flame that can't be seen in daylight; gasoline acts as a flame colorant, making methanol fires visible. The problem is that the components that would color a flame would burn off the quickest, leaving an invisible flame after a short pause.

The Big Three auto manufacturers are already well along in developing the first generation of production methanol cars, due in the 1993 model year. They know many of the pitfalls from their experience using ethanol, or grain alcohol, as a low-level additive in gasoline and, in Ford's case, as a pure fuel in Brazil. Meanwhile, the auto makers are testing thousands of cars, trucks, and buses converted to operate on methanol and placed in the fleets of state and local governments, particularly in California and New York. California has ordered 2,200 Chevrolet Lumina sedans that will operate on M85.

EPA Favors Alternatives

I wish I could tell you I felt methanol's brutal yet invigorating power in the pit of my stomach as I floored the accelerator of the M85 test car at GM's Technical Center. But first of all, Frank Ament, a GM engineer specializing in methanol as an automotive fuel, was behind the wheel, and, second, the Corsica seemed like any other, except for the electronic bar graph between the tachometer and speedometer. The display registered midway between eight and nine on a scale of 1 to 10, meaning, indeed, the fuel mix was M85. . . .

The EPA, which helped shape the Bush administration's policy, favors alternative fuels because the agency feels that the bulk of emissions improvement possible with gasoline engines has already been made.

Smog—the result of hydrocarbons reacting with nitrogen oxides in sunlight to produce highly reactive ozone and other harmful organic compounds—is the target of the latest round of regulations. Emissions of hydrocarbons, or partially burnt fuel, have been cut to four percent of what they were 20 years ago; the Bush plan calls for a further 50 percent reduction. Nitrogen oxides have been reduced to 24 percent of their level in 1970—a further cut of one-third is called for.

Air Quality

The environmental pollutant that has consistently resisted control is ground-level ozone. In 1987, 18 cities regularly exceeded federal air-quality standards for ozone, the reactive form of oxygen; Los Angeles was the worst of all, with a whopping 141 days of ozone alerts. Together with cuts in industrial emissions, the administration proposes to bring most cities into compliance with ozone standards by the beginning of the 21st century, though the worst offenders—Los Angeles, New York, and Houston—would probably need an extra decade to meet them.

Unfortunately, global warming, the environmental catastrophe that could potentially change the face of the Earth, is only slightly affected by the use of methanol as an alternative fuel. Of the alternative fuels under consideration, only hydrogen does not produce heat-trapping carbon dioxide when burned. Electric vehicles seem benign, but at least in the short term their power would come from coal- or oil-fired generating stations, so there is no gain on the greenhouse-gas front. The greater combustion efficiency of methanol cuts carbon dioxide output by about 10 percent. But this is a drop in the global bucket because most of the gases implicated in global warming come from industry and natural sources like swamps, volcanoes, and the methane produced by digestion and decomposition.

Still, if you make methanol from the natural gas now being flared at the wellhead, you get a net savings in carbon dioxide emissions, or, at minimum, better use of the natural gas. Though a good deal of carbon dioxide is released during the conversion of natural gas to methanol, the amount is no worse than that produced by refining gasoline from petroleum. Despite vast domestic coal resources, at present it appears pointless to make methanol from coal because far more carbon dioxide would be released at the refinery than is saved at the car's tailpipe. However, there are early indications that the carbon dioxide produced would be pure enough to be captured, compressed, and sold as a chemical commodity.

Will methanol actually improve the air we breathe? That is harder to answer than you might guess because the atmosphere is such a complicated system. Current models show that cutting hydrocarbon emissions by switching to methanol in Los Angeles would, over time, slightly reduce the city's notorious smog. But Joe Colucci, head of the fuel and lubrication department at GM's research laboratory, says the same scenario in another city wouldn't necessarily produce any benefit: "In Houston, where there is a large amount of hydrocarbon emissions, reducing it slightly with methanol cars isn't going to help."

"With natural gas resources so plentiful and widespread, its price could never be manipulated."

From an environmental perspective, methanol does have some distinct advantages over gasoline:

• Hydrocarbon emissions are lower. "Combustion is generally very complete, and when you take an engine apart that's been running on methanol, you see it's much cleaner," says GM's Ament. Methanol would cut tailpipe hydrocarbon emissions by between 20 and 50 percent right away, and by as much as 90 percent in advanced engines that can exploit its full potential. Further savings result from lowered evaporative losses during refueling and while the car is running, due to methanol's lower volatility. On top of that the methanol that does manage to escape is about half as reactive at forming smog as gasoline's components.

• Nitrogen oxides—which form when atmospheric nitrogen and oxygen combine at the high temperature and pressures in the engine's combustion chamber—can also be cut significantly because methanol burns cooler than gasoline.

"Methanol also promises to provide the United States with a second supply channel to balance against imported oil."

• Carbon monoxide will also fall because methanol can burn in an oxygen-rich environment.

On the negative side, methanol is a poison and an eye irritant, though the EPA believes dangerous levels could accumulate only under bizarre conditions—the fumes are just too easily dispersed. The real sticking point is formaldehyde emissions. This chemical was once commonly used to preserve biological specimens, but is now considered a carcinogen. About five times as much formaldehyde—a particularly efficient form of smog—is produced from methanol's combustion as from gasoline, most in the warmup phase. That could make it dangerous to breathe the air in unventilated garages.

The EPA's Hellman says that you could eliminate most of the formaldehyde if the car could be made to work as cleanly on starting as it does when fully warmed. "Cars now put out most of their pollutants in the first four minutes. The catalytic converter needs to be hot, and it needs the right mixture of gases to oxidize the emissions. In a rich burn it looks around and says, 'I'm hot, I'm ready—where's the oxygen?'"

One solution, he said, may be to inject fresh air into the catalyst during startup to improve the ratio of pollutants and oxygen. Another engineering fix would be to electrically heat the catalytic converter for about one minute before starting the engine, or place it closer to the engine. While these schemes would raise the efficiency of the converter during the early warmup stages, they could lead to a shorter life for the expensive pollution control device due to excessive heat build-up.

Another idea that has caught the fancy of both the oil industry and auto makers is changing the composition of gasoline to reduce the output of pollutants for older cars, which produce an inordinately large portion of air pollutants. This plan would reduce or eliminate the light components that vaporize easily and the extremely reactive heavy components. Says GM's Colucci, an advocate of what has come to be called reformulated gasoline: "Anything we can do to the current vehicles that are out there is a major plus."

The first . . . gasoline refiner to offer such a product is ARCO [Atlantic Richfield Company] with its Emission Control 1 gasoline (EC-l for short). . . . ARCO scientists say the gasoline will cut emissions from pre-1975 cars between four and nine percent. Comparable reductions from newer cars with catalytic converters would be harder and more costly with engines that have been optimized to run on current-grade gasoline. Colucci says that designing engines to run on reformulated gasoline would yield equivalent gains in emission reduction.

Blueprint for Efficiency

Besides the environmental advantages, methanol also promises to provide the United States with a second supply channel to balance against imported oil. Natural gas—presently the best candidate for producing methanol because of its clean process of conversion—is available in North America, South America, the Soviet Union, and in some Middle Eastern and Asian countries. With natural gas resources so plentiful and widespread, its price could never be manipulated. It would therefore tend to clip the wings of the Organization of Petroleum Exporting Countries, whose price hikes sent shock waves through the world economy in the 1970s.

The Big Three were already groaning under the burden of newly introduced pollution controls when they were caught off guard by expensive gasoline. This time it appears that the regulators have caught Detroit with its pants up. The auto makers have marshaled a huge contingent of chemists, material scientists, environmental engineers, and other experts whose lives revolve around efficient combustion, tailpipe catalysis, and effective maintenance of cars that run on alcohol. They are confident that they can handle whatever the government throws at them.

Alternative Fuels and Gasoline

But that doesn't stop their superiors in finance and accounting from worrying about the cost of alternate-fueled vehicles. It's clear that both the vehicle and its fuel would not compare favorably with gasoline. And it is certain that given the choice, consumers who own flexible-fuel vehicles would under the present pricing structure opt for gasoline unless the government provided incentives or subsidies for methanol.

Recent figures show an estimated cost of $22 billion for a methanol fleet large enough to displace 1,000,000 barrels of oil per day. The bulk of that cost would be spent on equipping cars to operate on methanol at an expected price tag of $300 per car.

As far as fuel costs go, a quantity of M85 with the energy equivalent of a gallon of gasoline has been forecast by the California Energy Commission to cost $1.44 in 1993. That's 23 cents a gallon more than premium unleaded gasoline is expected to cost, but that differential drops to pennies per gallon by the year 2000 when the bulk of methanol-powered cars are expected to be on the road, says the Commission's report.

Today, the largest producer of methanol is Hoechst-Celanese, with about one-tenth of the world's production of six billion gallons a year. At the present time there is plenty of methanol to go around. But at the levels planned for the turn of the century, some methanol may have to be diverted from the production of synthetic fibers and other industrial uses if new sources are not found and quickly brought on line.

"At the present time there is plenty of methanol to go around."

The oil companies have been quiet about what they might do, but it's a good guess that they would use their established infrastructure and marketing muscle to grab a big share of any market the government seems determined to create. Once the competitive juices start flowing, the price of methanol may come down.

Any new fuel or technology may seem impractical when viewed by skeptics who know enough about it to imagine what might go wrong. But you can really know its value only after you've given it a serious try. Ask yourself how gasoline would look if it were being proposed as an automotive fuel for the first time.

Electric Cars Can Replace Gasoline-Powered Cars

William O. Briggs Jr.

About the Author: *William O. Briggs Jr. has been employed as a technical consultant by major oil, gas, and coal companies. In 1979 Briggs founded the first electric-car dealership in California.*

Electric vehicles are not a futuristic pipe dream. They were the very first automobiles, and Thomas Edison himself bought electric car No. 2 off the Studebaker assembly line in 1902. This was not really surprising, because his company made the batteries that powered those first electric vehicles. The car had a top speed of 15 miles per hour, with a 50-mile range between charges. Between 1902 and 1911 Studebaker alone made about 2,000 electric cars and trucks.

However, due to range and speed limitations, electric vehicles (EVs) gradually gave up the market to internal combustion engine cars in the late 1920s. And so it remained (with few exceptions) until the gasoline crunches of the 1970s, which triggered a flurry of hurried manufacture and sales of electric vehicles ranging from glorified golf carts to vans. However, most of the companies involved were entrepreneurial start-ups which offered little service support after the sale, and most of these companies eventually dissolved. This left EV owners on their own to find parts and service. Moreover, the range of these vehicles was so short and inconsistent that they were impractical for almost every use. In any event, as soon as the gasoline lines disappeared, so did the EV market.

In contrast, the decade of the 1990s began with strong interest in electric vehicles being spurred by ongoing air quality concerns in more than 100 U.S. cities, as well as in many European cities. With transportation being a major contributor to urban air pollution, low-emission, alternative-fuel vehicles hold considerable promise for helping to relieve the problems of air pollution.

This has led to programs sponsored by both government and electric utilities. Federal government programs are under the auspices of the Department of Energy (DOE), with research and development being carried on through laboratories, universities, automobile companies, battery companies, and specialized engineering and component manufacturing companies.

On the state government level, the most notable agency involved is the California Energy Commission, with programs analogous to the DOE's. In January 1988, the California Electric Vehicle Task Force, made up of several public agencies and electric utility organizations, was formed with the purpose of facilitating the commercialization of EV technology.

In the private sector, the Electric Power Research Institute (EPRI), whose members are the major electric utility companies, is the primary agent for electric vehicle research and development.

"With transportation being a major contributor to urban air pollution, . . . alternative-fuel vehicles hold considerable promise."

Nationwide, the region experiencing the most serious air pollution is the Greater Los Angeles Basin. In response, the South Coast Air Quality Management District (SCAQMD)—consisting of Los Angeles, Orange, and Riverside counties, plus the nondesert portion of San Bernardino County—was created and empowered by enabling legislation in 1977. It has been

William O. Briggs Jr., "Electric Vehicles." This article appeared in the August 1990 issue and is reprinted with permission from *The World & I*, a publication of The Washington Times Corporation,

accorded incremental additional authority by a series of legislative amendments. Based on SCAQMD's initiative, it is likely that legislation enacted by September 1990 will mandate that owners of fleets of passenger and midsized internal combustion engine vehicles must convert them to the so-called ultra-low-emitting vehicles according to a fixed schedule. By July 1993, the required percentage of ultra-low-emitting vehicles is 2 percent. By 1996, the required percentage is to rise to 40 percent. From 1997 through 1999, this proportion rises to 75 percent, until the year 2000, at which time ultra-low-emitting vehicles would account for 100 percent of fleet replacements and additions. As of August 1990, the California Air Resources Board's research shows that electric vehicles and compressed natural gas vehicles are the only two types coming under ultra-low-emitting vehicles classification.

"Some 19.7 million EVs could be served nationwide with projected off-peak electricity-generating capacity."

Additionally, The Los Angeles Initiative, a broad coalition of community organizations, has been formed with the aim of marketing 10,000 EVs to commercial fleets in southern California by 1995. . . .

Response from van fleet owners and managers across the nation to a survey conducted by EPRI's Electric Vehicle Development Corporation shows that the market for EVs is ready to grow. In like manner, their electric utility survey respondents—representing approximately 17.3 percent of the total electricity sales in the United States—indicated that by the year 2000, 3.4 million EVs could be served by off-peak (nighttime) utility capacity. Extrapolating based on total electricity sales gives an estimate that some 19.7 million EVs could be served nationwide with projected off-peak electricity-generating capacity. In comparison, the Motor Vehicle Manufacturers Association estimates that by the year 2000 a total of approximately 210 million cars, trucks, and buses will be operating on the U.S. roads. Thus, the projected electrical generating capacity of the United States by the year 2000 could service approximately 9 percent of the estimated total U.S. motor vehicles by that year.

Electric Vehicle Technology

Given the interest in the market by potential users, and the capacity and readiness of the electric utility industry, where does EV technology stand today? The answer appears to be "on the threshold," with a likely near-term market realization of surprising breadth.

Introducing substantial numbers of electric vehicles will require the development not only of an efficient electric vehicle but also of a complex electric fueling network, just as there is today a complex network of gas fueling stations. Thus, a review of the prospects for introducing electric vehicles must mention several different technologies.

EVs use electricity stored in rechargeable batteries to power electronically controlled direct current or alternating current motors. Lead-acid batteries are the old technology—durable and reliable, tried and proven—but they have a relatively low energy and power density compared with some other batteries that either now are or soon will be commercially available. An improved lead-acid battery system, designated as "HED 88" (high-energy density) powers the GM [General Motors] Griffon—a British-made electric van that has already logged over seven million miles in fleets around the world. For the North American market, the proven technology of the Griffon is the foundation of the G-Van, which was developed by EPRI and will be manufactured by the VEHMA Corporation of Canada.

The most probable next generation of battery is the nickel-iron battery technology developed by Eagle Picher Industries of Joplin, Missouri. The Chrysler Electric TE-Van—using their standard minivan design—is scheduled to use this nickel-iron battery system. Tests have indicated a range of 120 miles and a top speed of 65 miles per hour.

Following closely to the nickel-iron is the sodium-sulphur battery technology. ASEA Brown Boveri of Switzerland and Sweden is one of the prime movers in this regard, with successful testing using Volkswagen Golf model cars (called CitySTROMers) in West Germany. Similar successful testing is being done by Chloride Silent Power, Ltd., of England. Moreover, sodium-sulphur has been selected as the battery system of choice for the Ford ETX-II Aerostar Electric Van. The sodium-sulphur battery holds high promise for long range (200 miles) and life (exceeding five years).

Other battery technologies currently in advanced research and development are (a) nickel-cadmium, (b) zinc-bromine, (c) lithium aluminum-iron sulfide, (d) sodium/metal chloride, (e) iron-air, and (f) aluminum-air, each with a different degree of promise for contributing to longer range, longer life, and higher traveling speeds for EVs.

Roadway Power

An alternative to the "pure" EV is the electric hybrid vehicle (EHV), which incorporates an on-board auxiliary power unit—a small multifuel engine/generator set—in the event the battery has insufficient charge to run the car. The auxiliary power unit is a range extender and power enhancer, which will reduce concern about power and range, while batteries, as the primary power unit, reduce air pollution emissions from the routine combustion of fuels. The EHV thus offers a transitional "bridge" from fossil fuel-powered vehicles to "pure" EVs.

With partial funding from the California Energy Commission, the Electric Auto Association of Belmont, California, is developing an EHV using a 35-horsepower rotary engine, with a Chevrolet Corsica frame and body housing it, the battery pack, and the components.

Meanwhile—"down the road" further—but not that far down —comes the exciting "roadway powered electric vehicles," which derive most or all of their power for propulsion from electric sources buried in conventional-looking roadways. EVs traveling along the roadway extract electrical energy from an energized source through the process of magnetic induction. The energy may be used either to power the vehicles directly or to charge on-board batteries used for travels off the electrified roadway.

The unique elements of the roadway-powered technology that are not addressed in the conventional EV development programs are the vehicle-mounted inductive pickup element and the core elements embedded in the roadway.

On April 18, 1990, Southern California Edison Company and the Los Angeles Department of Water and Power announced the joint launching of the Playa Vista Roadway Powered Electric Vehicle Demonstration Project, consisting of a 5,000-foot section of resurfaced roadway which includes a 1,000-foot test section of powered roadway. Test vehicles will include two G-Vans and the roadway-powered passenger bus developed in the Program in Advanced Technology for the Highway at the University of California at Berkeley.

This demonstration project is the precursor to a network of electric roadways consisting of selected lanes of streets and freeways throughout southern California, with projected start-up implementation by the mid-1990s, leading to the culmination of a completed wide-reaching "grid" by the year 2000.

"EVs traveling along the roadway extract electrical energy from an energized source through . . . magnetic induction."

Two additional technologies, both with exciting potential applications for EVs, are fuel cells and solar cells. The thermoelectrochemical fuel cell converts methanol fuel or heat into electrical power for propulsion and is presently undergoing a two-year research and development program by the DOE. Thin-film amorphous silicon photovoltaic solar cells—which convert sunlight directly into electricity and can be deposited on many surfaces, including metal, glass, ceramic, and plastic—are a promising EV power source, especially in Sun Belt areas.

"Electric vehicles are 98 percent less polluting than gasoline-powered cars per mile traveled."

Given a viable power source, the EV then needs efficient delivery of the power to the motor. For EVs with powertrains using an alternating current motor (or motors), power "inverters" have been developed which convert the direct current from batteries to an alternating current. These and other components of the EV, such as the motor itself, are in an advanced state of development and manufacture.

Having a viable battery pack, the next task is to achieve recharging in a convenient and reliable mode. The obvious such mode is the "overnight" charge, by plugging in the EV at home, office, or wherever and getting a full recharge in six to eight hours. This also fits the electric utilities' capacity by utilizing their off-peak hours.

However, given the limited range (60-120 miles) between full recharges, other modes become necessary to enhance EV use. The Inductran Corporation of Berkeley, California, has developed both the technology for rapid partial charging of an EV through magnetic coupling at a nonmoving station and an alternative philosophy—called "opportunity charging"—for implementing large-scale EV use. Using Inductran's basic battery-charging technology, an EV could drive up to a charging station and via magnetic coupling receive either a partial or full recharge. The Opportunity Charging concept proposes that by making recharging fast and easy, it will become possible to achieve acceptable performance levels in EVs with smaller battery packs, and the batteries will last longer.

Among several technologies claiming to provide safe recharging in 30 minutes or less is one developed by Electronic Power Devices Corporation of Atlanta. Their innovative charging technology uses many cycles of "jolting" a battery with high current for a few hundred milliseconds, then discharging it for two or three milliseconds. This technology has successfully recharged the batteries in the video cameras at Turner Broadcasting Systems, Inc. in Atlanta, and is successfully operating in several major companies.

And the electric utilities have indicated that they can and will provide off-street and/or curbside charging pedestals operated much like parking meters. Notwithstanding that these types of chargers would likely be used during peak-load times, Southern California Edison has indicated that multiple thousands of EVs thus "plugging in" could be accommodated with their present generating capacity.

Another option under consideration is the placement of battery exchange stations along major routes to enable an EV user to exchange his discharged battery pack for a fully charged one, thus providing a much extended range.

Upcoming Plans

As with any new product, start-up production and construction of infrastructure for EVs represent higher costs than the expense of maintaining such a system. Taking into account base vehicle cost, conversion cost, battery cost, fuel price, fuel economy, maintenance, insurance, operating characteristics, interest rates, and salvage value, the expense of EV transportation is probably comparable to today's gasoline-powered vehicles. Example: A comparison of a conventional van with an electric van shows that total life-cycle costs over an eight-year period are about equal when electricity is priced at five cents per kilowatt hour and gasoline at $1.15 per gallon.

The South Coast Air Quality Management District has estimated that for every 100,000 miles driven, gasoline-powered passenger cars emit 2,623 pounds of reactive organic gases, 2,574 pounds of carbon monoxide, and 172 pounds of nitrogen oxides. For the same distance, pollution associated with EVs—including generation—totals but 5, 9, and 61 pounds, respectively. In total, electric vehicles are 98 percent less polluting than gasoline-powered cars per mile traveled.

On April 18, 1990, General Motors announced that they are proceeding with their plan to produce and sell their "Impact" EV, which had been introduced at the Los Angeles Auto Show in January. The sleek subcompact coupe is powered by 32 10-volt sealed lead-acid batteries and uses two AC induction motors. It can accelerate from 0 to 60 miles per hour in a phenomenal 8.0 seconds and achieves a 120-mile range at a cruising speed of 55 miles per hour. GM has thus given an immense boost to EV reality and Impact may well set the standard other manufacturers will attempt to match or exceed.

Fiat Auto of Italy began marketing their "Panda Elettra" EV in June 1990, offering a cruising range of 75 miles with top speed of 70 miles per hour with lead-acid batteries, with a range increase to 110 miles with optional nickel-cadmium batteries. In 1990 Fiat planned to sell 500 of these at a cost of $20,000 each. In France, meanwhile, both Peugeot and Renault announced plans to market two-passenger light-duty EVs during 1990, Peugeot using nickel-cadmium and Renault using nickel-iron batteries.

Aside from the automakers, the Electric Vehicle Association of the Americas, a nonprofit corporation headquartered in Silver Spring, Maryland, has been formed with the purpose of promoting all aspects of electric and hybrid vehicle development. . . .

EVs have crossed the threshold into reality, and the decade of the 1990s promises to see them consolidate their position as a primary technology for the twenty-first century.

Bicycles Can Replace Gasoline-Powered Cars

Marcia D. Lowe

About the Author: *Marcia D. Lowe researches alternatives in agriculture and transportation for the Worldwatch Institute, a Washington, D.C. organization that analyzes global resource and environmental problems.*

In a world so transformed by the automobile that whole landscapes and lifestyles bear its imprint, a significant fact goes unnoticed. While societies the world over define transportation in terms of engine power, the greatest share of personal transport needs is met by human power.

From the 10-speeds of Boston to the black roadsters of Beijing, the world's 800 million bicycles outnumber cars by two to one—and each year bike production outpaces automobile manufacturing by three to one. Bicycles in Asia alone transport more people than do all of the world's autos.

In developing countries—where urban workers cycle to their jobs and rural dwellers pedal two- and three-wheelers piled high with loads of goods—pedal power is an important part of national economies and the only alternative to walking that many people can afford. In industrial countries, bicycles are a practical supplement to motorized transport. . . .

A comparison of bicycle and automobile ownership illustrates each country's reliance on bicycles for actual transport. The United States has more than seven times as many bicycles per person as India, but because 1 out of every 2 Americans owns an automobile—compared with 1 out of 500 Indians—bicycles play a much more modest role in the U.S. transportation system.

Western Europeans are the industrial world's heaviest bicycle users. In several countries—among them Denmark, West Germany, and the Netherlands—bike owners outnumber nonowners. Pro-bicycle planning in the last twenty years (rather than conducive climate or flat terrain, as is often assumed) distinguishes Europe's truly "bicycle-friendly" countries. The Netherlands and Denmark lead this group, with bicycle travel making up 20 to 30 percent of all urban trips—up to half in some towns. In several European countries, 10 to 55 percent of railway patrons in suburbs and smaller towns arrive at the station by bicycle. Traffic jams and air pollution in the past decade have spurred authorities in Switzerland, West Germany, and Austria to encourage more bicycle use. Cyclists in the United Kingdom, Belgium, France, and southern European countries, by contrast, still enjoy little support.

"In industrial countries, bicycles are a practical supplement to motorized transport."

Many automobile-saturated cities of North America and Australia have all but abandoned the bicycle for the automobile. The growth of suburbs has sent jobs, homes, and services sprawling over long distances that inhibit cycling, as well as mass transit. Many major cities are largely bicycle-proof, their roadways and parking facilities designed with only motor vehicles in mind. While Australia has a set of planning and design guidelines for bicycle transportation, state and local governments often ignore them, lacking a sense of commitment to transport options other than the automobile. Several North American cities explicitly emphasize safety for cycle commuters—including Seattle, Calgary, and some university towns—but these are exceptions. . . .

The automobile—which has brought indus-

Marcia D. Lowe, "The Bicycle: Vehicle for a Small Planet," *Worldwatch Paper 90*, September 1989. Reprinted with permission of the Worldwatch Institute.

trial society a degree of individual mobility and convenience not known before—has long been considered the vehicle of the future. But countries that have become dependent on the car are paying a terrible price: each year brings a heavier toll from road accidents, air pollution, urban congestion, and oil bills. Today people who choose to drive rather than walk or cycle a short distance do so not merely for convenience, but also to insulate themselves from the harshness of a street ruled by the motor vehicle. The broadening of transport options beyond those that require an engine can help restore the environment and human health—indeed, the very quality of urban life.

Car Accidents

Despite improvements in safety, an estimated quarter of a million people worldwide die in automobile accidents each year. Millions more are injured. Developing countries—with fewer automobiles but more pedestrian traffic and no provisions for separating the two—have fatality rates per vehicle-mile up to 20 times higher than industrial countries, and road accidents have become a leading cause of death. . . .

Excessive motorization deepens the oil dependence that is draining national economies. In 1988, U.S. oil imports cost some $26 billion, or more than 20 percent of the country's foreign trade deficit. Many nations' transport sectors account for more than half of their petroleum consumption. In the United States, the share is 63 percent, and in Kenya 91 percent. Even with the recent fall in crude prices, fuel bills are particularly harsh for indebted Third World countries that spend large portions of their foreign exchange earnings on imported oil. In 1985, low-income developing countries (excluding China) spent on average 33 percent of the money earned through merchandise exports on energy imports; many spent more than half.

The oil shocks of the seventies brought home the precarious nature of petroleum as an energy source, exposing car-dependent countries' vulnerability. Today's stable prices and supply have lulled importers into complacency. But as oil demand continues to rise, another oil crisis looms in the nineties.

"Excessive motorization deepens the oil dependence that is draining national economies."

The world's largest gridlocked cities may run out of momentum before they run out of oil. Traffic in many cases is moving at a pace slower than bicycles, with average road speeds during peak times sometimes down to 8 kilometers per hour. Some police units in European and North American cities—including London, Los Angeles, Victoria, and many others—use bicycles rather than squad cars for patrolling congested urban centers.

In some parts of the Third World, urban gridlock is even worse. Traffic density in Taiwan's capital is 10 times that of congestion-plagued Los Angeles, and work trips take Mexico City commuters up to four hours each day. In Nigeria's capital—which experienced a doubling of its fleet during the oil boom of the seventies—"one quarter of the problem of working on Lagos Island is getting to work," according to one civil servant. "The rest is getting home."

The economic and social costs of congestion, already daunting, are bound to multiply if car commuting trends persist. The Confederation of British Industry warns that traffic congestion costs Britain $24 billion a year—including employee time lost through tardiness, and inflated goods prices resulting from higher distribution costs. The U.S. Federal Highway Administration put the country's loss to traffic jams at $9 billion in 1984. The FHWA expects a fivefold increase in that amount by 2005. Some 50 percent more cars are projected to be on the road then, the typical commuter's 10- or 15-minute delay may stretch to an hour, and roads will likely be congested throughout the day.

To date, planners have typically sought only technological solutions to auto-induced problems. Without support for alternatives to driving, however, these fixes may be inadequate. For example, the catalytic converter that has contributed to impressive reductions in hydrocarbon and carbon monoxide emissions in the United States actually slightly increases the carbon dioxide buildup that contributes to climate change. And while the converter and other technological cures have dramatically reduced pollution from U.S. passenger cars since the early sixties, rapid growth in the vehicle fleet and miles traveled have partially offset this progress.

Similarly, the quest for petroleum alternatives has focused largely on "clean" fuels such as methanol, made from coal or natural gas, and alcohol substitutes distilled from corn and other crops. But methanol contributes to ozone formation and, if derived from coal, to climate change by emitting twice as much carbon dioxide per unit of energy as does gasoline. Using crop-based feedstocks for alternative fuels poses its own environmental side-effects, and a potential conflict with food production.

No Room for Roads

Nor is building more roads the answer to congestion. Transport planners are finding that constructing new freeways just attracts more cars, as some public transit riders switch to driving and new developments spring up along the new roads. In 1988, a California Department of Transportation study concluded that neither a $61 billion road building program, nor *any* further road building, could solve its traffic problem.

Even if such solutions could work, society is running out of resources to devote to the automobile. Industrial-world cities typically relinquish at least one-third of their land to motor vehicles for roads and parking lots. As former FHWA administrator Robert Farris said: "We can no longer completely build our way out of the congestion crisis by laying more concrete and asphalt. Time is too short, money is too scarce, and land is often not available."

Third World countries with mounting pressures to house and feed their swelling populations have even less room to spare for private automobiles. Where people and good cropland are concentrated in a relatively small area, as in China, choices are few. For China to pave over as much land per capita as has the United States (about .06 hectares) would mean giving up a total of 64 million hectares—equivalent to more than 40 percent of the country's cropland.

"Bicycle transportation could help restore balance to people whose daily lives are in many ways governed by the automobile."

For every person who makes a trip by bicycle instead of by car there is less pollution, less fuel used, and less space taken on the road. . . .

A 1980 study in Great Britain calculated that if just 10 percent of car trips under 10 miles were made by bicycle, the country would save 14 million barrels of oil a year, or 2 percent of total consumption. In 1986, a national campaign in the Netherlands encouraged drivers to switch to bicycles for trips within a 5-kilometer radius of home. Policymakers estimated that this would save each motorist at least $400 a year in fuel costs.

A 1983 study of American commuters revealed that just getting to public transit by bicycle instead of by car would save each commuter roughly 150 gallons of gasoline a year. When a motorist who otherwise drives all the way to work switches to bike-and-ride, his or her annual gasoline use drops by some 400 gallons, half the amount consumed by the typical car in a year. At current prices, if 10 percent of the Americans who commute by car switched to bike-and-ride, nearly $1 billion would be shaved off the U.S. oil import bill.

Bicycle transportation also uses space more efficiently than automobile transport. In fact, all other modes, especially mass transit options, can move more people per hour in a lane of a given size than an automobile even at the top of its range. Since bicycles offer the same degree of individual mobility as automobiles, replacing short car trips with bicycling could ease congestion without curbing people's freedom to move when and where they choose. Replacing longer car trips with mass transit—especially with cycling as the way to get to the station—would save even more space on roads.

Relief from traffic is something everyone can enjoy, not just drivers who choose to cycle instead. Many cities in the industrial world temporarily close major streets to motorized traffic, in effect turning them into wide cycle and foot paths. In 1983, Mayor Augusto Ramirez Ocampo launched such a program in Bogotá, Colombia, under the slogan "the city for the citizens." Every Sunday morning 56 kilometers of arterial roads are closed to motor traffic and half a million city dwellers take to the streets to cycle, roller skate, or stroll.

Bicycle transportation could help restore balance to people whose daily lives are in many ways governed by the automobile. Philosopher Ivan Illich has concluded that the average American male "spends four of his sixteen waking hours [driving his car] or gathering his resources for it."

"The bicycle is truly a vehicle for a small planet."

The daily battle with traffic congestion, according to a recent University of California study, tends to raise drivers' blood pressure, lower their frustration tolerance, and foster negative moods and aggressive driving. Except when there is no alternative but to ride in the same traffic stream, commuter cyclists benefit both themselves and their employers by being less vulnerable to hypertension, heart attacks, and coronary disease, and by arriving at work more alert. The proof that people enjoy cycling to keep fit is in the popularity of stationary exercise bikes; the irony, however, is that so many people drive to a health club to ride them.

Reassured by its benefits both to society and the individual, the bicycle is truly a vehicle for a small planet. "Bicycling is human-scale," writes New York cycling activist Charles Komanoff. "Bicycling remains one of New York City's few robust ecological expressions . . . a living, breathing alternative to the city's domination by motor vehicles. There is magic in blending with traffic, feeling the wind in one's face, the sheer fact of traversing the city under one's own power."...

Bicycling Communities

North America's closest approach to a cycling society is the bicycle-friendly university town. Two such communities in northern California, Palo Alto and Davis, vie for the title of bicycling capital of the United States. Davis has the higher cycling rate of the two—25 percent of total trips in the community of 44,000 are made by bike—and cycle trailers filled with groceries or children are not an unusual sight. Davis has some 30 miles of bicycle lanes for 100 miles of streets, and roughly 20 miles of separate cycle paths.

Palo Alto, an affluent, highly educated community of 56,000 near San Francisco, has gone beyond physical improvements to promote bicycle transportation. The city government pays its employees seven cents a mile for all business travel by bicycle, and sponsors a city-wide monthly "Leave Your Car at Home Day." Palo Alto's police department has a bicycle squad. The city has a traffic school for juveniles who violate bicycle laws and funds an on-road cycling course for middle school students.

Palo Alto has spent roughly $1 million since 1980—mostly from state grants—on bicycle lockers and racks, bike bridges, and lighted cycle paths. All road patching in town must adhere to high smoothness standards. Bicycle-detecting traffic sensors are clearly marked so that cyclists can easily activate them. The centerpiece of Palo Alto's 40-mile system of bikeways is its bicycle boulevard, a two-mile stretch in the middle of town where bikes are the only through traffic allowed. The bicycle boulevard is just the first segment of a larger network planned for the city center.

A 1983 zoning ordinance requires new buildings beyond a certain size to provide secure bicycle parking and showers for employees. Several large employers in Palo Alto add their own incentives. The Alza Corporation, for example, pays bicycle commuters $1 for each day they ride to work. Amenities at Xerox—which include a towel service in the shower room—help explain why 20 percent of the company's local employees cycle to work, one of the highest bicycle commuter rates nationwide.

In many parts of the United States, public support for cycling facilities has suffered from a deep divide among cycling advocates. On one side are those who oppose any special cycleways or routes that are separate from motor vehicle traffic, believing them unsafe and discriminatory. On the other side are cyclists who advocate building special bikeways where practical.

This division has undermined official cycling promotion, since many local authorities dismiss requests for new bikeways on the grounds that cyclists themselves do not want them. Meanwhile, millions of people annually use the 2,700 miles of abandoned rail corridors that have been converted into cycling and hiking trails largely through the efforts of the Rails-To-Trails Conservancy—a private organization.

A Growing Trend

Scattered but thriving bicycle programs outside the main cycling societies show a growing trend toward pro-bicycle planning. Many towns in Sweden, Switzerland, and West Germany have steadily increasing shares of cyclists in their traffic, and national and local authorities have stepped up their commitment to cycle planning in recent years. West Germany's years of *Verkehrsberuhigung*, or traffic calming, have helped foster a cycling environment by restraining motorized traffic with physical barriers and reduced speed limits. Swedish towns have experimented with restraining motor vehicles since the seventies by using "traffic cells" that divide a city into zones to reroute traffic from denser areas onto main roads, making smaller streets safer for cycling and walking.

"A cycling society can emerge out of one already hooked on the automobile."

Several cities in Canada and Australia are drafting or implementing cycling plans. Montreal—a city with four to five months of hard winter—plans to double its present 200 kilometers of bikeways by the year 1993, aiming to have a cycle lane or path within 2 kilometers of every point on the city's street system. Melbourne, Perth, and other Australian cities are following the example of the city of Geelong's 1977 bicycle plan, which aims to make every street bikable.

There is ample proof that a cycling society can emerge out of one already hooked on the automobile. Palo Alto council member Ellen Fletcher, who cycles to city meetings and has become known nationally for her bicycle advocacy, knows how to fulfill the bicycle's potential. "All you have to do is make it easier to ride a bike than drive a car," she says. "People will take it from there."

Bibliography

Books

Michael Brower — *Cool Energy*. Cambridge, MA: Union of Concerned Scientists, 1990.

John Byrne and Daniel Rich, eds. — *Planning for Changing Energy Conditions*. New Brunswick, NJ: Transaction Books, 1988.

Bernard L. Cohen — *The Nuclear Energy Option: An Alternative for the 1990s*. New York: Plenum Publishing Co., 1990.

Art Davidson — *In the Wake of the Exxon Valdez: The Devastating Impact of the Alaska Oil Spill*. San Francisco: Sierra Club Books, 1990.

Edward R. Fried and Nanette M. Blandin, eds. — *Oil and America's Security*. Washington, DC: Brookings Institution, 1988.

Jose Goldemberg et al. — *Energy for a Sustainable World*. New York: Wiley, 1988.

Lawrence Gregory Hines — *The Market, Energy, and the Environment*. Boston: Allyn and Bacon, 1988.

Frederick Uhlen Hop — *The Energy-Saving House Design Handbook*. Englewood Cliffs, NJ: Prentice Hall, 1989.

Joseph G. Morone and Edward J. Woodhouse — *The Demise of Nuclear Energy?: Lessons for Democratic Control of Technology*. New Haven, CT: Yale University Press, 1989.

Raymond L. Murray — *Nuclear Energy: An Introduction to the Concepts, Systems, and Applications of Nuclear Processes*. 3rd ed. Oxford: Pergamon Press, 1988.

Stanley M. Nealey — *Nuclear Power Development: Prospects in the 1990s*. Columbus, OH: Battelle Press, 1990.

Roberta J. Nichols — *Transportation Fuels for the Future: Why Alternative Fuels?* Wichita, KS: Wichita State University, College of Engineering, 1990.

Dixy Lee Ray and Lou Guzzo — *Trashing the Planet: How Science Can Help Us Deal with Acid Rain, Depletion of the Ozone, and Nuclear Waste (Among Other Things)*. Washington, DC: Regnery Gateway, 1990.

Periodicals

M.A. Adelman — "Oil Fallacies," *Foreign Policy*, Spring 1991.

Carla Atkinson — "America's Energy Future," *Public Citizen* May/June 1991.

Ed Begley Jr. — "My Generation," *Omni*, May 1991.

Sharon Begley — "Alternative Energy: Time to Get Serious," *Newsweek*, August 20, 1990.

Sharon Begley — "The Power of a Voltswagon," *Newsweek*, April 1, 1991.

Sharon Begley — "Running on Swamp Gas," *Newsweek*, September 10, 1990.

Peter A.A. Berle — "Mobilizing Energy Efficiency," *Audubon*, March 1991.

Jan Beyea — "Nuclear as Last Resort to Change in Climate," *Forum for Applied Research and Public Policy*, Fall 1990. Available from Bruce Brocka, Executive Sciences Institute, 1005 Mississippi Ave., Davenport, IA 52803.

Bill Breen — "Burn It?" *Garbage*, March/April 1991. Available from Old House Journal Corp., 435 Ninth St., Brooklyn, NY 11215.

Corie Brown — "Utilities Are Making More by Selling Less," *Business Week*, January 9, 1989.

Lester R. Brown, Christopher Flavin, and Sandra Postel — "A Global Plan to Save Our Planet's Environment," *USA Today*, January 1990.

Craig Canine — "The Second Coming of Energy

Conservation," *Utne Reader,* January/February 1990.

Betsy Carpenter — "A Nuclear Graveyard," *U.S. News & World Report,* March 18, 1991.

Ginny Carroll — "The Furnace Beneath Us," *Newsweek,* August 28, 1989.

Sarah Chasis and Lisa Speer — "How to Avoid Another Valdez," *The New York Times,* May 20, 1989.

Jeremy Cherfas — "Skeptics and Visionaries Examine Energy Saving," *Science,* January 11, 1991.

James R. Chiles — "Tomorrow's Energy Today," *Audubon,* January 1990.

Lee Clarke — "Oil-Spill Fantasies," *The Atlantic,* November 1990.

Barry Commoner — "Making Solar Top Priority," *Mother Jones,* March/April 1991.

William J. Cook — "Jump-Start to the Future," *U.S. News & World Report,* April 30, 1990.

Milton R. Copulos — "Curing Our Import Habit," *The World & I,* December 1990.

Robert W. Crandall and John D. Graham — "New Fuel-Economy Standards?" *The American Enterprise,* March/April 1991. Available from the American Enterprise Institute for Public Policy Research, 1150 17th St. NW, Washington, DC 20036.

Philip C. Cruver — "Lighting the Twenty-First Century," *The Futurist,* January/February 1989.

Abe Dane — "The Hydrogen Age," *Popular Mechanics,* September 1990.

John Dillin — "Greater Use of Wind, Solar Urged," *The Christian Science Monitor,* March 14, 1991.

Gregg Easterbrook — "Tilting at Windmills," *The New Republic,* March 18, 1991.

George Everett — "Energy-Saving Facts and Fables," *Consumers' Research Magazine,* January 1990.

Arthur Fisher — "Next Generation Nuclear Reactors: Dare We Build Them?" *Popular Science,* April 1990.

Michael Fumento — "Smog and Mirrors," *The Wilson Quarterly,* Winter 1991.

Michael Fumento — "What Kind of Fuel Am I?" *The American Spectator,* November 1990.

The Futurist — "Energy Efficiency," September/October 1988.

Michael W. Golay — "Longer Life for Nuclear Plants," *Technology Review,* May/June 1990.

Michael W. Golay and Neil E. Todreas — "Advanced Light-Water Reactors," *Scientific American,* April 1990.

Christine Gorman — "Mid-Life Crisis for Nukes," *Time,* March 18, 1991.

Peter F. Gray — "Kicking the Oil Habit," *The Washington Monthly,* March 1991.

Wolf Hafele — "Energy from Nuclear Power," *Scientific American,* September 1990.

Kent Hansen et al. — "Making Nuclear Power Work," *Technology Review,* February/March 1989.

Eric Hirst — "Boosting U.S. Energy Efficiency," *Environment,* March 1991.

Eric Hirst — "Demand-Side Management," *Environment,* January/February 1990.

Eric Hirst — "Electricity: Getting More With Less," *Technology Review,* July 1990.

Holman Jenkins Jr. — "Oil Marketplace Freedom Staving Off Big Shortages," *Insight,* August 27, 1990.

Mary Kadlecek — "Alternative Energy—Hot Again?" *The Conservationist,* January/February 1991.

Paul Klebnikov — "Demand-Side Economics," *Forbes,* April 3, 1989.

Tom Knudson — "Rancho Seco Decked Again," *The Bulletin of the Atomic Scientists,* December 1989.

Daniel E. Koshler Jr. — "Solar Power and Priorities," *Science,* August 25, 1989.

Lynette Lamb — "Making Fossils of Fossil Fuels," *Utne Reader,* May/June 1991.

Amory B. Lovins — "Save It, Don't Burn It," *Mother Jones,* March/April 1991.

Francesca Lyman — "Rethinking Our Transportation Future," *E Magazine,* September/October 1990. Available from Earth Action Network, 28 Knight St., Norwalk, CT 06851.

Robert Malpas — "Moving Toward Greater Energy Efficiency," *Scientific American,* September 1990.

Alan S. Miller and Irving M. Mintzer — "Global Warming: No Nuclear Quick Fix," *The Bulletin of the Atomic Scientists,* June 1990.

Anne S. Moffat — "Methanol-Powered Cars Get Ready to Hit the Road," *Science,* February 1, 1991.

Frank H. Murkowski — "America Needs Alaska's Oil," *The New York Times,* April 1, 1991.

National Review — "Seabrook A-borning," April 30, 1990.

Christina Nichols — "Renewable Energy: Lay Your

Money Down," *Technology Review*, July 1990.

Peter Nulty — "Oil's Prospects: A Better Decade," *Fortune*, April 22, 1991.

Bill Paul — "U.S. Is Rapidly Losing Its Lead in Alternative Energy," *The Wall Street Journal*, August 15, 1989.

John E. Petersen — "How to Kick the Gasoline Habit," *Governing*, March 1991.

Alan Reynolds — "Malign Neglect," *Forbes*, December 10, 1990.

Andrei Sakharov — "Mankind Cannot Do Without Nuclear Power," *Time*, May 21, 1990.

Scott Saleska — "Low-Level Radioactive Waste: Gamma Rays in the Garbage," *The Bulletin of the Atomic Scientists*, April 1990.

John C. Sawhill — "The Energy/Environment Interface," *Vital Speeches of the Day*, December 1, 1990.

Science — "Growth Without New Energy," June 22, 1990.

Scientific American — "Energy for Planet Earth," September 1990.

Phil Scott — "Good to the Last Drop," *Omni*, May 1991.

Peter H. Stone — "You're Energy Czar, Now What?" *Mother Jones*, March/April 1991.

John J. Taylor — "Improved and Safer Nuclear Power," *Science*, April 21, 1989.

John Templeman — "Fill 'Er Up,—With Hydrogen, Please," *Business Week*, March 4, 1991.

Brian Tokar — "Energy Blues and Oil," *Z Magazine*, January 1991.

Utne Reader — "Radioactive Waste: As Close as Your Neighborhood Landfill," September/October 1990.

Sara van Dyck — "Wind at Work," *Sierra*, November/December 1990.

Francoise Vaysse — "Reviving Nuclear Power," *World Press Review*, January 1991.

Harvey Wasserman — "Bush's Pro-Nuke Energy Strategy," *The Nation*, May 20, 1991.

Carol J. Weinber and Robert H. Williams — "Energy from the Sun," *Scientific American*, September 1990.

Woody West — "Toward the Land of Ecotopia," *Insight*, September 24, 1990.

Tim Wise — "Blood for Oil?" *Dollars & Sense*, May 1991.

Richard Woodbury — "Bright Hopes for the Blue Flame," *Time*, September 24, 1990.

Richard Woodbury — "How to Break the Middle East Oil Habit," *Time*, October 29, 1990.

David Woodruff — "GM Drives the Electric Car Closer to Reality," *Business Week*, May 14, 1990.

David Woodruff — "Is America Finally Ready for the Gasless Carriage?" *Business Week*, April 8, 1991.

Daniel Yergin — "How to Design a New 'Energy Strategy,'" *Newsweek*, February 11, 1991.

Organizations to Contact

The editors have compiled the following list of organizations that are concerned with the issues debated in this book. All of them have publications or information available for interested readers. The descriptions are derived from materials provided by the organizations. This list was compiled upon the date of publication. Names and phone numbers of organizations are subject to change.

American Petroleum Institute (API)
1220 L St. NW
Washington, DC 20005
(202) 682-8000

The American Petroleum Institute is a trade association representing America's petroleum industry. Its activities include lobbying, conducting research, and setting technical standards for the petroleum industry. API publishes numerous position papers, reports, and information sheets.

American Solar Energy Society (ASES)
2400 Central Ave., Suite B-1
Boulder, CO 80301
(303) 443-3130

ASES promotes solar energy. It disseminates information on solar energy to schools, universities, and the community. In addition to the *ASES Publications Catalog,* the society publishes the bimonthly magazine *Solar Today.*

American Wind Energy Association (AWEA)
1730 N. Lynn St., Suite 610
Arlington, VA 22209
(202) 408-8988

AWEA supports wind energy as an alternative to current energy sources. It provides federal and state legislators with information on wind as an energy source. Publications include the *AWEA Wind Energy Weekly* and the monthly *Windletter.*

Competitive Enterprise Institute (CEI)
233 Pennsylvania Ave. SE, Suite 200
Washington, DC 20003
(202) 547-1010

CEI advocates removing government environmental regulations to establish a system in which the private sector is responsible for environment and energy policy. Its publications include the monthly newsletter *CEI UpDate* and numerous reprints and briefs.

Conservation and Renewable Energy Inquiry and Referral Service (CAREIRS)
PO Box 8900
Silver Spring, MD 20907
(800) 523-2929

CAREIRS, a project of the U.S. Department of Energy, answers inquiries and provides referrals and information concerning the use of renewable energy technologies and conservation methods. CAREIRS publishes fact sheets, including *Renewable Energy: An Overview, Wind Energy,* and *Solar Energy and Your Home: Questions and Answers.*

Council on Alternate Fuels (CAF)
1225 I St. NW, Suite 320
Washington, DC 20005
(202) 898-0711

CAF is comprised of companies interested in the production of synthetic fuels and the research and development of synthetic fuel technology. It publishes information on new alternative fuels in the monthly *Alternate Fuel News.*

Energy Conservation Coalition (ECC)
1525 New Hampshire Ave. NW
Washington, DC 20036
(202) 745-4874

ECC is a group of public interest organizations that promote energy sufficiency. It supports government policies that encourage energy conservation. ECC publishes *Powerline,* a bimonthly periodical covering consumer issues on energy and utilities.

Geothermal Resources Council (GRC)
PO Box 1350
Davis, CA 95617
(916) 758-2360

GRC encourages the research, exploration, and development of geothermal energy, and supports legislation and regulations that promote the use of such energy. It provides information to the public and publishes the monthly *GRC Bulletin.*

Greenpeace
1436 U St. NW
Washington, DC 20009
(202) 462-1177

Greenpeace opposes nuclear energy and supports ocean and wildlife preservation. The organization sponsors public protests against nuclear energy and other activities it believes harm the environment. It publishes the bimonthly magazine *Greenpeace* and many books and reports, including *Global Warming: The Greenpeace Report* and *Saving Energy Is Saving Lives.*

The Heritage Foundation
214 Massachusetts Ave. NE
Washington, DC 20002
(202) 546-4400

The Heritage Foundation is a public policy think tank that advocates that the United States increase domestic oil production. Its publications include the quarterly magazine *Policy Review,* brief *Executive Memorandum* editorials, and the longer *Backgrounder* studies.

International Association for Hydrogen Energy (IAHE)
PO Box 248266
Coral Gables, Fl 33124
(305) 284-4666

The IAHE is a group of scientists and engineers professionally involved with the production and use of hydrogen. It sponsors international forums to further its goal of creating an energy system based on hydrogen. The IAHE publishes the monthly *International Journal of Hydrogen Energy.*

National Coal Association (NCA)
1130 Seventeenth St. NW
Washington, DC 20036
(202) 463-2625

NCA is a national trade association that represents the coal industry. The association is primarily a lobbying organization that advocates the use of coal to meet America's energy needs. It publishes the weekly *Coal News* newsletter, the bimonthly magazine *Coal Voice,* and periodic forecasts, fact sheets, and monographs.

Natural Resources Defense Council (NRDC)
40 W. Twentieth St.
New York NY 10011
(212) 727-2700

The council is a nonprofit activist group composed of scientists, lawyers, and citizens who work to promote environmentally safe energy sources and protection of the environment. NRDC publishes a quarterly, *The Amicus Journal,* the newsletter *Newsline,* and a bibliography of books concerning environmental issues.

Political Economy Research Center (PERC)
502 S. Nineteenth Ave., Suite 211
Bozeman, MT 59715
(406) 587-9591

PERC is a research and education foundation that focuses primarily on environmental and natural resource issues. Its approach emphasizes the use of the free market and the importance of private property rights in protecting the environment and finding new energy resources. Publications include *PERC Viewpoint* and *PERC Reports.*

Renewable Fuels Association (RFA)
1 Massachusetts Ave. NW, Suite 820
Washington, DC 20001
(202) 289-3835

RFA is comprised of professionals who research, produce, and market renewable fuels, especially alcohol fuels. It also represents the renewable fuels industry before the federal government. RFA publishes the monthly newsletter *Ethanol Report.*

Union of Concerned Scientists (UCS)
26 Church St.
Cambridge, MA 02138
(617) 547-5552

The Union of Concerned Scientists is an organization of scientists and other citizens concerned about nuclear energy and the impact of advanced technology on society. The UCS conducts independent research, sponsors and participates in conferences and panels, and testifies at congressional and regulatory hearings. The UCS publishes a quarterly newsletter, *Nucleus,* as well as books, reports, and briefing papers.

United States Council for Energy Awareness (USCEA)
1776 I St. NW
Washington, DC 20006-2495
(202) 293-0770

The United States Council for Energy Awareness is the public relations organization for the commercial nuclear energy industry in the United States. Its activities include public and media relations and public opinion research. USCEA's publications include the monthly newsletter *INFO,* the quarterly *Nuclear Industry* magazine, and position papers, fact sheets, and brochures.

Worldwatch Institute
1776 Massachusetts Ave. NW
Washington, DC 20036
(202) 452-1999

The Worldwatch Institute is a nonprofit research organization that analyzes and focuses attention on global problems, including environmental and energy concerns. The institute, which is funded by private foundations and United Nations organizations, publishes the bimonthly *World Watch* magazine and the Worldwatch Paper Series, including *Air Pollution, Beyond the Petroleum Age; Designing a Solar Economy; Alternatives to the Automobile: Transport for Livable Cities,* and *Slowing Global Warming: A Worldwide Strategy.*

Index